Water Residuals to Reduce Soil Phosphorus

Awwa Research Foundation

About the Awwa Research Foundation

The Awwa Research Foundation (AwwaRF) is a member-supported, international, nonprofit organization that sponsors research to enable water utilities, public health agencies, and other professionals to provide safe and affordable drinking water to consumers.

The Foundation's mission is to advance the science of water to improve the quality of life. To achieve this mission, the Foundation sponsors studies on all aspects of drinking water, including supply and resources, treatment, monitoring and analysis, distribution, management, and health effects. Funding for research is provided primarily by subscription payments from approximately 1,000 utilities, consulting firms, and manufacturers in North America and abroad. Additional funding comes from collaborative partnerships with other national and international organizations, allowing for resources to be leveraged, expertise to be shared, and broad-based knowledge to be developed and disseminated. Government funding serves as a third source of research dollars.

From its headquarters in Denver, Colorado, the Foundation's staff directs and supports the efforts of more than 800 volunteers who serve on the board of trustees and various committees. These volunteers represent many facets of the water industry, and contribute their expertise to select and monitor research studies that benefit the entire drinking water community.

The results of research are disseminated through a number of channels, including reports, the Web site, conferences, and periodicals.

For subscribers, the Foundation serves as a cooperative program in which water suppliers unite to pool their resources. By applying Foundation research findings, these water suppliers can save substantial costs and stay on the leading edge of drinking water science and technology. Since its inception, AwwaRF has supplied the water community with more than $300 million in applied research.

More information about the Foundation and how to become a subscriber is available on the Web at **www.awwarf.org.**

Water Residuals to Reduce Soil Phosphorus

Prepared by:
James DeWolfe
McGuire Environmental Consultants, Inc.
712 Hampshire Circle, State College, PA 16803

Jointly sponsored by:
Awwa Research Foundation
6666 West Quincy Avenue, Denver, CO 80235-3098

Pennsylvania-American Water Company

Tulsa Metropolitan Utility Authority

and

Denver Water

Published by:

Awwa Research Foundation

American Water Works Association

IWA Publishing

DISCLAIMER

This study was jointly funded by the Awwa Research Foundation (AwwaRF) and the co-sponsoring utilities. AwwaRF and the utilities assume no responsibility for the content of the research study reported in this publication or for the opinions or statements of fact expressed in the report. The mention of trade names for commercial products does not represent or imply the approval or endorsement of AwwaRF or the utilities. This report is presented solely for informational purposes.

CONTENTS

TABLES

FIGURES

FOREWORD

The Awwa Research Foundation is a nonprofit corporation that is dedicated to the implementation of a research effort to help utilities respond to regulatory requirements and traditional high-priority concerns of the industry. The research agenda is developed through a process of consultation with subscribers and drinking water professionals. Under the umbrella of a Strategic Research Plan, the Research Advisory council prioritizes the suggested projects based upon current and future needs, applicability, and past work. The recommendations are forwarded to the Board of Trustees for final selection. The foundation also sponsors research projects through the unsolicited proposal process; the Collaborative Research, Research Applications, and Tailored Collaboration program; and various joint research efforts with organizations such as the U.S. Environmental Protection Agency, the U.S. Bureau of Reclamation, and the Association of California Water Agencies.

This publication is a result of one of these sponsored studies, and it is hoped that its findings will be applied in communities throughout the world. The following report serves not only as a means of communicating the results of the water industry's centralized research program but also as a tool to enlist the further support of the nonmember utilities and individuals.

Projects are managed closely from their inception to the final report by the foundation's staff and large cadre of volunteers who willingly contribute their time and expertise. The foundation serves a planning and management function and awards contracts to other institutions such as water utilities, universities, and engineering firms. The funding for this research effort comes primarily from the Subscription Program, through which water utilities subscribe to the research program and make an annual payment proportionate to the volume of water they deliver and consultants and manufacturers subscribe based on their annual billings. The program offers a cost-effective and fair method for funding research in the public interest.

A broad spectrum of water supply issues is addressed by the foundation's research agenda: resources, treatment and operations, distribution and storage, water quality and analysis, toxicology, economics, and management. The ultimate purpose of the coordinated effort is to assist water suppliers to provide the highest possible quality of water economically and reliably. The true benefits are realized when the results are implemented at the utility level. The foundation's trustees are pleased to offer this publication as a contribution toward that end.

Walter J. Bishop
Chair, Board of Trustees
Awwa Research Foundation

Robert C. Renner, P.E.
Executive Director
Awwa Research Foundation

ACKNOWLEDGMENTS

The author of this report is indebted to the sponsoring water treatment utilities and their staff for financial and field support to conduct this Tailored Collaboration project. The sponsoring utilities and their staff that supported this project include:

- Pennsylvania-American Water Company
  - Randy Pankiewicz
  - Paul Zielinski
- Tulsa Metropolitan Utility Authority (OK)
  - Brian Bolte
  - Shawn Glen
  - Ray West
  - Roy Foster
  - Tom Krueger
- Denver Water (CO)
  - Ken Lykens
  - Tom Mountfort

Invaluable research was conducted by Nicholas Basta, PhD, Associate Professor of Soil and Environmental Chemistry and Elizabeth Dayton, PhD, Research Associate, both of The Ohio State University, School of Natural Resources, 2021 Coffey Road, 210 Kottman Hall, Columbus, OH.

Special recognition is given to: Jennifer Weld and Andrew Sharpley, PhD of the United States Department of Agriculture – Agriculture Research Service (USDA-ARS) Pasture Systems and Watershed Management Research Unit located at The Pennsylvania State University, University Park, PA; James Ippolito, PhD of Colorado State University, and; Larry Poindextor of the USDA Natural Resources Conservation Service (NRCS) of Oklahoma. Their insight into nutrient management programs and comments on the benefits of soil amendments are greatly appreciated.

Heartfelt appreciation is offered to Joe Bernosky, Nicole Graziano, Chad Seidel of McGuire Environmental Consultants, Inc. for their assistance in compiling data from the project and generating the final report.

The authors would also like to specially thank Albert Ilges and India Williams, AwwaRF Project Managers, and the Project Advisory Committee (PAC) members for their guidance, suggestions, and patience throughout this study. PAC members included:

- James (Jim) Parsons, Director of Engineering, Cobb County-Marietta Water Authority (GA)
- Jeff Novak, USDA-ARS, Coastal Plains Soil, Water and Plant Research Center (SC)

A special tribute and dedication is given to Bill Stout, a PAC member who passed away during the conduct of this study. Bill was a pioneer in the use of materials to control phosphorus loss from soils. His inspiration to colleagues and his insight for beneficial use of residuals were both special contributions to this project.

EXECUTIVE SUMMARY

BACKGROUND

This research report is the result of a tailored collaboration project for several utilities that have unique, yet common, interests in managing water treatment residuals (WTRs) generated by their drinking water treatment facilities. Their concerns are unique because of the differing chemical characteristics of their respective WTRs as well as the individual state regulations regarding the management of these materials. Their commonality of interests resides with their concerns about source water quality and the impacts that watershed activities – specifically activities related to agricultural practices – could have on their source water qualities. The co-sponsors for this research included:

- Pennsylvania American Water Company (PAWC)
- Tulsa Metropolitan Utility Authority (TMUA - Tulsa, OK)
- Denver Water (DW - Colorado)

There is also the mutual desire on the part of the co-sponsors to further develop an understanding of how WTR management can be linked to source water management, particularly within their respective watersheds.

There are historic concerns regarding the impact of phosphorus (P) from agricultural lands on source water quality – concerns that have increased with the need to secure additional water resources. Each of the co-sponsoring utilities has experienced source water challenges due to non-point source (NPS) P pollution, either directly in the waters withdrawn from the sources or within impoundments/reservoirs where water from the sources is stored prior to treatment.

Each of the co-sponsoring utilities also has experienced challenges when managing WTRs for a variety of reasons including, but not limited to, 1) processing (dewatering), 2) storage space, or 3) their ultimate fate (reuse or disposal). Thus, establishing beneficial use programs in which the WTRs generated at the drinking water treatment facilities are used for source water protection would create a "win-win" situation, whereby WTR could be used in an environmentally responsible manner within a utility's watershed and have direct benefit on source water quality.

As part of a national initiative, National Resource Conservation Service (NRCS) state offices across the United States, along with other state and federal agencies, have been developing programs to better manage NPS P pollution. These programs control fertilizer, animal manure, and biosolids applications based on state-specific guidelines that consider numerous factors. Some states have incorporated the use of best management practices to reduce the risk of P pollution.

OBJECTIVE

The goal of the co-sponsoring utilities, and the focus of this research project, is to exploit the unique characteristics of WTR for beneficial use. Specifically, the targeted beneficial use is improved P management of agricultural lands by controlling the release of P. The mechanism for this approach is to use metal hydroxides present in WTR to bind P in agricultural soils that have high P levels. Beneficial use could be realized by using WTR as a best management practice (BMP) to reduce soluble P runoff from agricultural land and allow for quantitative reductions in P pollution. The use of WTR can serve as an effective BMP in states that use a P Index approach for managing agricultural lands that receive large amounts of animal manure from concentrated animal feeding operations (CAFOs). Also, the P binding capacity of soils can be increased due to the addition of metal hydroxides (described in Chapter 2). The P Index approach is being used by state regulating agencies to consider numerous factors related to the potential sources and transport of P from agricultural lands to nearby waterways, and is explained in detail in Chapter 1. This report is intended for general water treatment industry dissemination and use. Specifically, the project was designed to:

- Evaluate how different WTRs reduce P loss from agricultural land and lower P Index scores.
- Develop recommendations and land application procedures for beneficial use of WTRs based on chemical characteristics of different types of WTR.
- Develop tailored recommendations for beneficial use of WTRs as BMPs.

APPROACH

To investigate the beneficial use of WTRs to control NPS P runoff and protect surface water quality, a research program was developed that evaluated several key items.

- Several types of WTRs were selected and characterized based on the coagulant used (aluminum and iron based) and their form – dewatered or liquid.
  The ability to remove dissolved P from runoff-water and reduce P solubility in soil or organic materials was evaljuated. Different loading rates were evaluated to determine if incremental impacts of loading would occur. Tests were performed over time to gage the impacts of WTR application. This was accomplished through
  - Blending WTRs with high P soils
  - Blending WTRs with a biosolid and poultry manure
  - Applying WTRs to a soil surface and simulating rainfall.
- Correlate the results of WTR characterization with test results to determine if any quantitative analyses could be correlated to determine general trends and to guide WTR selection for beneficial use to control NPS P pollution.

CONCLUSIONS

Extended Characterization Value

The results of extended characterization (discussed in detail in Chapter 3) demonstrated that each WTR has unique chemical characteristics that impact the ability of each WTR to react with P. These unique characteristics are attributable to the following:

- Type of coagulant used
- Dose of the coagulant
- Source water constituents

- Direct impacts of chemical properties of treatment chemicals
- Indirect chemical properties of treatment chemicals
- Storage conditions
- Age of water treatment residuals

There are several tests that provided valuable insight into the ability of WTR to bind P, which correlated well with research results, the most significant being inorganic-P sorption maximum values (Pmax). There was a high degree of correlation between Pmax and the concentration of reactive aluminum and iron as measured by ammonium oxalate extraction. The tests also revealed that several WTRs possessed substantial levels of major soil nutrients, including nitrogen and potassium. The presence of these nutrients is likely due to source water conditions; however, the presence of nitrogen could also be attributed to the use of amine based polymers as coagulant, flocculant, filter, or dewatering aids.

P Reduction When Incorporated in High P Soils

Tests evaluating the blending of WTR with a high P soil (313 mg/kg via Mehlich 3 analysis) demonstrated that loading rates between 1 and 10% (0.2 and 2.2 Mg/ha) were able to significantly reduce both soil test P and soluble P. The following conclusions were possible based on the test results:

- All WTRs had varying degrees of impact on both soil test P and soluble P. This helps to defend the general thesis that each WTR possesses unique P adsorption characteristics.
- All WTRs were able to reduce soil test P to less than 200 mg/kg at the 10% loading rate, and in most cases were able to do this after just one week of incubation time.
- The majority of the soil test P reduction occurred after only one week of incubation. This is likely due to the large surface area available on the solids surfaces of both the WTRs and the high P soil.
- Of the WTRs tested, the most reactive was an iron coagulant-based WTR; however, general assumptions should not be made about the relative sorption capacity of iron based-coagulants compared to aluminum based-coagulants, since only one iron coagulant based WTR is included in this comparison.
- WTR that is stored in lagoons and directly land applied without dewatering prior to application exhibited substantially greater ability to bind P when compared to the same WTR that had been dewatered. This difference is likely due to the more amorphous and reactive nature of the aluminum hydroxides (less crystalline and mineralized) when they are in the liquid form, along with the likelihood that the wet materials have a greater surface area than when they are dried.
- The maximum amount of liquid WTR that can be applied to a soil will likely be related to the hydraulic capacity of the soil such that runoff does not occur from the application. Significant correlations were observed between Pmax and the percent reduction in both soil test P and soluble P. This helps to confirm the value of this parameter for guiding the selection of WTRs for beneficial use as a P binding soil amendment.

P Reduction in Runoff When Surface Applied

Simulated rainfall testing also confirmed the ability of WTRs to reduce P in runoff for soils that are high in P and have also recently received organic material (poultry manure). The addition of liquid WTRs caused a decrease in runoff P. Significant observations include:

- The WTR applied as liquid reduced P slightly better than when dried. This is likely attributable to liquid WTRs having a greater reactive oxide surface area than when dried, which is consistent with observations made in the high P soil blending tests.
- As with the incubations results, strong correlations (R^2 values of 0.52 to 0.92) exist between the Pmax values determined during the initial characterization and the results of the simulated runoff tests for the six solid WTRs.
- Other factors including storage costs, dewatering costs, distance to use, differential performance between wet and dry WTR, and transportation costs must all be considered when evaluating residuals management options.
- The majority of P runoff from the test plots occurred after the initial rainfall event. This further confirms the potential benefit of co-applying WTR at the same time organic material (in this case, chicken manure) is applied (to gain the greatest benefit of its P sorption potential).
- The research evaluated WTR loaded in a concentrated strip (25% of surface area). It is reasonable to assume there is a strong potential that WTR loaded on greater areas of agronomic fields, or incorporated with manure prior to application, could further reduce soluble P.

P Reduction When Blended with Organic Materials

Tests that blended WTR with biosolids and poultry manure revealed that WTRs were able to substantially reduce both soil test P and soluble P, and that the percent reduction was directly related to the amount of WTR used. Generally, soluble P levels were substantially reduced by all WTRs at loading rates at or above 25% ratio (dry weight basis, WTR to organic material). Also, most WTRs exhibited little or no incremental increased impact on soluble P above a 25% loading. It should be noted that substantially high blending ratios (WTR:organic material) would require large amounts of WTR, which may not be practical from an availability perspective; however, the intent of WTR addition is not to reduce the fertilizer value of organic materials, but rather to control the rate at which P is released – especially after initial application.

RECOMMENDATIONS

Water Treatment Residual Characterization

Utilities should maintain ongoing and current characterization of their respective WTRs. This characterization should include all the parameters specified by the state regulatory agencies that oversee the management of WTR, together with parameters that determine the ability of the WTR to bind P. These parameters include Pmax, as determined through isotherm testing, and reactive aluminum and iron, as determined through an ammonium oxalate extraction or similar methods. Currently, only research-level laboratories possess the ability to conduct isotherm testing to develop Pmax values that this research showed to be highly predictive of their ability to reduce P pollution potential. Therefore, each utility is encouraged to identify and establish routine dialogues with collaborators that are developing or have established these advanced testing methods, or other methods that could also more easily determine the maximum P sorption potential of WTR.

Advanced analyses can be coupled with routine testing that is required by state regulatory agencies. The differences in WTRs observed during this research support the recommendation that utilities characterize their WTRs on a seasonal basis due to changes in source water quality or the required level of pretreatment chemicals. California-based utilities should confirm these results using the California Waste Extraction Test, a more aggressive leaching procedure that is used in that state to characterize hazardous waste.

Communicate Results to Nutrient Management and Regulatory Agencies

Each of the co-sponsors, and other drinking water utilities that are interested in using WTRs as a BMP for soil P pollution control should actively pursue relationships with local USDA-ARS and NRCS offices and related agencies to communicate the potential role of WTR in managing P in agronomic soils and organic materials management programs, especially those involved with the management of CAFOs. Utilities should also communicate the results of this research to state agencies that are responsible for regulating the management and fate of WTR. This is especially true for states that have emerging WTR regulations, WTR regulations that are being reviewed, or established beneficial use programs.

Pennsylvania

Existing Beneficial Use permits should be reviewed to determine if P control can be adopted within the conditions, and to determine if a completely new general permit specific to soil P management is warranted. Further, Pennsylvania's biosolids regulations should be reviewed and appropriate dialogues established to determine the role WTR can play in the control of P for the land application of biosolids and manures.

Other state agencies, including the Pennsylvania Department of Environmental Protection, Bureau of Watershed Management, Pennsylvania Association of Conservation Districts (PACD), and Chesapeake Bay Foundation should be contacted to determine funding opportunities to demonstrate full-scale use of WTR for NPS P pollution control.

Colorado

Colorado has recently revised their regulations for WTR management, administered through the Colorado Department of Public Health and Environment (CDPHE) Water Quality Control Division. This regulation does contain provisions for beneficial use, which require the submittal of a Beneficial Use Plan. The co-application of WTR with biosolids is regulated under Colorado's Biosolids Regulations; therefore the conditions of these regulations also should be reviewed.

The presence of radionuclides is an emerging issue for WTR management in Colorado, prompted by the Radionuclides Rule (USEPA 2000) and the release of draft guidance to address the management of WTRs that may contain radionuclides. Dialogue with state regulators is encouraged to utilize the results of this research to show that substantial environmental benefits can be gained through the direct land application of WTR, or co-application with biosolids, and that the loading rates utilized may not have a substantial impact on background radiation levels of unamended soils.

Oklahoma

Oklahoma does not have formal WTR regulations, but has been at the forefront of activities to address P pollution in watersheds recognizing their substantial impacts on water quality in drinking water supplies. In fact, one of the two demonstration watersheds used nationally for implementing federal nutrient management requirements was the Lake Eucha Watershed. The degraded water quality within the watershed and Lake Eucha has been directly linked to the runoff of P-laden materials from densely concentrated poultry-growing operations in the watershed. Oklahoma utilities, and especially TMUA, should communicate the results of this research to lawmakers that are finalizing regulations for the management of organic materials (manures) from CAFOs.

Utilize NRCS Programs and Codes

BMPs that utilize WTR for NPS P pollution must control the delivery of P to receiving water resources by:

- Minimizing P availability (source reduction),
- Reducing the transport and/or delivery of P, and
- Intercepting P in runoff before it is delivered to the water resource.

Utilities should identify the potential role of WTR in their state's NRCS Conservation Practice Standards that have been developed to guide activities related to agronomic practice. A detailed list of standards that could utilize WTR are provided in Chapter 5. It is very important to note that while application of WTRs to sequester P in high P soils or organic byproducts must be associated with a reduction in future application rates of P (in manure or biosolids). Adding WTRs does have BMP potential, but is a stop gap measure until input reductions are achieved.

WTR Processing

The results of this research showed that liquid WTRs – those stored in lagoons or similar structures – are more reactive for binding P when compared to the same material after it has been dried. Therefore, utilities that use lagoons should consider this fact when considering the ultimate fate of WTR from storage units when they are cleaned to provide capacity for continued operation. The dewatered WTRs used in this research were substantially processed to help normalize the particle size of WTR; however, the actual particle size of WTR used in land application will vary substantially between utilities. This will likely reduce the rate of P adsorption, and programs that use WTR for soil P control should use appropriate field monitoring methods to carefully monitor the impact WTR has on P binding within and P leaching from soils.

Further, utilities that are considering modifying WTR management programs for greater capacity should include the enhanced ability of liquid WTR to bind P as a selection factor. Since watershed activity has a major influence on source water quality and the required treatment to accomplish finished water goals, watershed protection should be pursued to the extent possible.

Future Research

The results produced from this tailored collaboration project identified several items that require further investigation to further define the role WTR can play as a BMP for the control of P loss from agronomic soils to protect water quality.

- Controlled field demonstrations should be conducted to better translate the impact of WTR for soil P control. The field demonstrations could occur on a small or large scale, depending on the available resources. The larger the scale of the demonstration, a greater number of samples will have to be acquired.
- Development of a simple predictive model that can be used to assess the potential of a given WTR for P control. The model input would only require typical chemical parameters to characterize the WTR eliminating the need for further laboratory analysis of the residuals.
- Defining clear and simple recommendations for handling, storing, and transporting WTR that minimize management costs yet maximize soil P control characteristics.
- $CaCl_2$ extractable P was used in this research to represent the release of P within a soil solution and thus, P that may pass through a soil in soil solution. Water extractable soil P is becoming widely accepted as representing the release of soil P to rainfall and runoff water, and thus, the transfer of soil P to runoff. Therefore, soluble P via both these methods should be compared fro WTR and soil/organic byproduct blends to determine the best method to gage the impact of WTR on bioavailable P.
- This project focused on P loss due to the magnitude of impact that P loss has on surface water quality. There may be other constituents worthy of measurement in runoff waters that would address other potential public health concerns, including but not limited to nitrogen, arsenic, chromium, copper, and zinc.
- While WTR can bind P in-situ or in runoff water, the fate of WTR bound P that enters the aquatic environment with particulates due to erosion should be evaluated.
- Plant uptake or accumulation of other WTR constituents via plant tissue analyses and crop yields should be documented at field scale trials to further assess the impact of WTR on the soil matrix and plants.
- The longevity of the effect of WTRs on soils also should be considered to determine if WTRs have to be reapplied after a certain number of years to accomplish the same effect on soil P. Such evaluation would include an assessment of the transformation of P between inorganic and organic forms.
- This research was conducted with one soil type to limit the variables for determining the impact of WTR on both soil test and soluble P. Additional studies on other soils types also are recommended to identify which soil parameters make the greatest difference in the ability of WTR to have short term and long term effects on soil P.
- The impact of WTR on soil biotic properties (i.e., soil fungi) should be examined, since P transformation can be substantially impacted by these properties.

CHAPTER 1
BACKGROUND AND INTRODUCTION

OVERVIEW

Water utilities are increasingly concerned about source water quality. This concern is reflected in both recent and future regulations as well as the managerial and operational concept of total water management. Although aimed primarily at reducing the risk of *Cryptosporidium* and other microbial pathogens in drinking water, the Long Term 2 Enhanced Surface Water Treatment Rule toolbox (USEPA 2003) specifically cites implementing (or improving) a Watershed Control Program as part of the multi-barrier approach for treating drinking water.

Under the Clean Water Act (CWA), the concept of Total Maximum Daily Loads (TMDLs) is attempting to address whole water body quality. As the TMDL program advances and matures, regulatory agencies are focusing on NPS water pollution in addition to the more traditional (and theoretically easier to identify and control) point sources (USEPA 1997).

NPS pollution, unlike point source pollution from industrial and sewage treatment plants, comes from many diffuse sources. NPS pollution is generally attributed to rainfall or snowmelt moving over and through the ground. As the runoff moves, it picks up and carries away natural and human-made pollutants, finally depositing them into lakes, rivers, wetlands, coastal waters, and even underground sources of drinking water. Major sources of NPS pollutants include fertilizers, herbicides, and insecticides from agricultural lands and residential areas, as well as a variety of contaminants in runoff from impervious (urban) areas (Leeds, Brown, and Watermeier 1992).

There are several other important operational issues that drinking water utilities must address in addition to treatment challenges associated with degradation of source water quality. One of these issues is the management of water treatment residuals (WTRs) that are generated during the water treatment process. WTRs commonly consist of constituents removed from the source water and solids generated from the addition of treatment chemicals. A major component of WTR is clay to silt-sized particles that contain metal hydroxides, which form after the addition of coagulants.

The ultimate fate of WTR is becoming an ever increasing concern. Traditional disposal methods have included landfilling, but liability issues associated with mixing of materials from other sources and higher disposal costs have prompted many utilities to search for alternatives. In many cases, the unique chemical and physical characteristics of WTR have prompted research to discover beneficial uses of these materials that historically have been viewed as wastes.

The co-sponsoring utilities for this research project – Pennsylvania American Water Company, Tulsa Metropolitan Utility Authority, and Denver Water – are very aware of the importance of source water management and have extensive experience in WTR management for beneficial use. It is their desire to further develop the understanding of how WTR management can be linked to source water management, particularly within the watersheds from which their facilities obtain source water.

The goal of the co-sponsoring utilities, and the focus of this research project, is to exploit the unique characteristics of WTR for beneficial use. Specifically, the targeted beneficial use is improved nutrient management of agricultural lands by controlling the release of P from agricultural soils, which may also lead to managing the quality of sources of water supplies.

SIGNIFICANCE OF PHOSPHORUS POLLUTION

One significant NPS contributing to receiving water bodies is agricultural run-off, which can be a source of microbiological, organic, and inorganic contaminants. Of the inorganic contaminants, P is the compound most often implicated in the degradation of surface water quality. P is an essential plant nutrient and considered to be the primary "limiting" (or controlling) growth factor for aquatic flora such as algae (Sharpley 2000). The introduction of high concentrations of P into a water body could cause accelerated and massive algal and aquatic weed growth. Excess algal growth can lead to a number of potentially significant adverse water quality impacts including eutrophication and algae blooms. Algae blooms alone can produce a number of undesirable consequences including:

- Taste and odor events (especially from the compounds geosmin and methylisoborneol (MIB)
- Water treatment plant filter interference (via clogging);
- Contribute organic carbon and exacerbate disinfection byproduct formation, and;
- Production of potentially toxic metabolic compounds.

Phosphorus Loss Mechanisms

Through the mechanisms of rainfall, irrigation, erosion, and subsequent runoff, P can be transported to nearby receiving water bodies such as rivers, lakes, and reservoirs. There are several pathways for P loss to occur from agricultural soils, depending on the form of P. The major portion (75-90%) is transported as particulate P as it attaches (via a number of mechanisms) to mineral and organic sediment. The immediate availability of particulate P for biological activity is variable and typically limited, however this particulate load can represent a long-term P source for eventual use by aquatic biota. Soluble P is of particular concern, due to it being readily available for biological activity. P loss also occurs when it is eroded from the soil surface by overland flow. A small portion of P may leach through soil and be present in water percolating through the soil profile. This is of particular concern where manure application has occurred for a substantial period of time (Novak et al. 2000). Usually, P levels are low in this fraction due to fixation on soil particles; however, sandy soils with low fixation or holding capacities can transfer substantial P. Also, soils with the preferential flow of water due to macropores and earthworm holes can rapidly transfer P bearing water (Sharpley and Syers 1979; Sims, Simard, and Joern 1998). Figure 1.1 illustrates the potential pathways for P loss from agricultural soil.

Sources of Phosphorus

A variety of animal manures are frequently used to fertilize cultivated lands with the primary objective being nitrogen (N) addition. The application process tends to be N-based, evaluating only soil N content and crop N requirements. Depending on these factors and the type of animal manure used this approach can frequently lead to excess P addition, often three times the nutrient requirement of the crop. Table 1.1 shows that different manures can contribute different levels of both N and P, so the source of manure must also be taken into consideration.

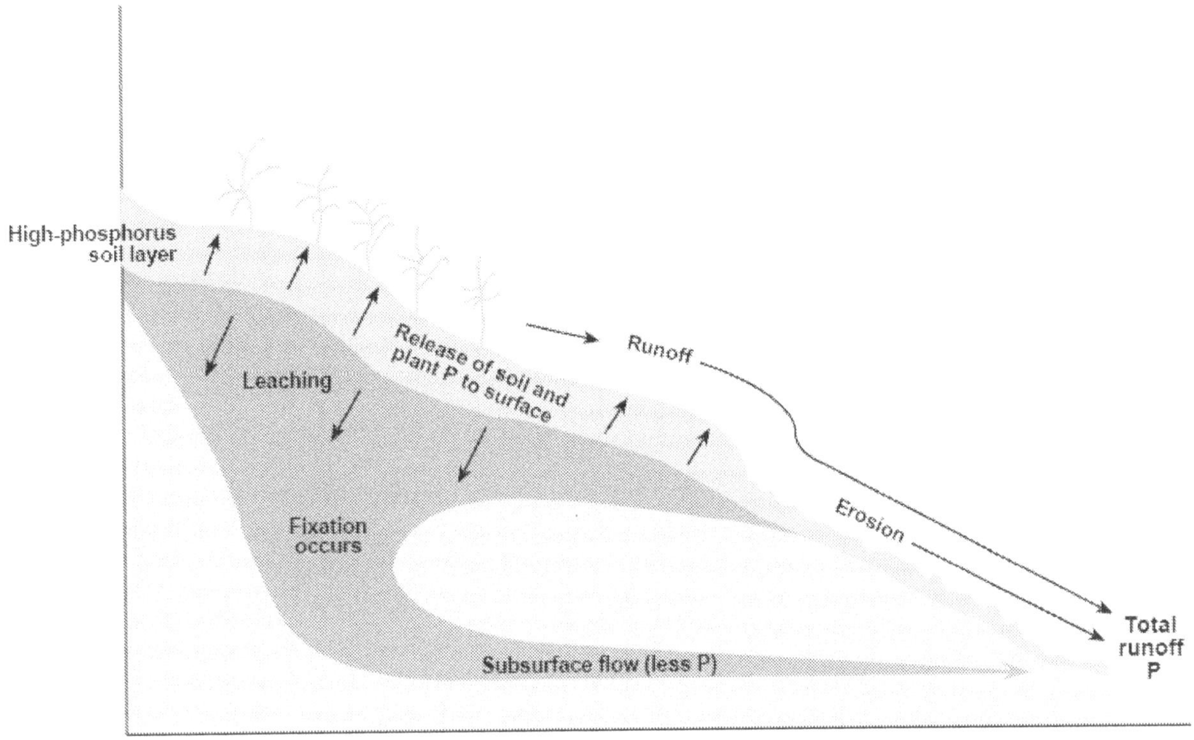

Source: Sharpley and Beegle 2001

Figure 1.1 Forms of NPS phosphorus loss from soil

Table 1.1
Average nitrogen and phosphorus content of animal manures

Animal	Units in lbs/day/1000 lb of animal		
	Total Manure	**Nitrogen**	**Phosphorus**
Beef	59.1	0.31	0.11
Dairy (lactating cow)	80.0	0.45	0.07
Hogs and pigs	63.1	0.42	0.16
Chickens (layers)	60.5	0.83	0.31
Chickens (broilers)	80.0	1.10	0.34
Turkeys	43.6	0.74	0.28

Source: Adapted from USDA-NRCS 1999

Inorganic P fertilizers are also commonly used, especially when initiating crop growth. All of these materials are highly water soluble, although solubility can be controlled through formulation. The analyses of P fertilizers are given as phosphorus oxide (P_2O_5).

Programs to Reduce Phosphorus Loss

Recognizing this potential threat to surface water quality, both the federal government and a number of states have developed guidance to control the non-point introduction of P into surface waters. A USDA sponsored program emerged in the early 1990's to develop assessment tools for areas with water quality problems, termed Southern Extension Research Activity (SERA) – Information Exchange Group 17. This group (the 17th formed under the auspices of this regional branch of the Cooperative States Research, Education, and Extension Service (CSREES), a USDA agency) consists of scientists from universities and governmental agencies.

SERA 17 participants recognized that while soil test P is related to P concentrations of runoff, different amounts of P can be lost from sites with similar surface soil test P contents. Prior studies showed that site characteristics such as slope and vegetative cover can have an overriding effect on P loss in runoff and erosion. Thus, P loss from fields with similar soil test P values can vary greatly as a function of soil management practices and any climatic, topographic, and agronomic factors that affect runoff and erosion. Thus, a more comprehensive approach than just a soil P test was needed for reliable, yet flexible recommendations of the environmental impact of P management. This need led to the development of the Phosphorus (P) Index.

Lemunyon and Gilbert (1993) described the P Index as an improved approach that accounts for transport potential and as a tool to rank potential threat of P loss from agricultural land to surface water. Since its introduction, the P Index has evolved into many versions, but the basic concept is the backbone of all approaches. Figure 1.2 illustrates the concept that provides a method to assess potential risk of P losses from a field based on P transport and management factors. Weighted scores are assigned to transport and management terms that affect potential for P runoff. Agricultural land with high P Index scores may be restricted from P additions (manure, biosolids, fertilizer) until the value is reduced.

The P Index has now been adopted by most states as one of the methods for controlling soil P losses in accordance with USDA-National Resource Conservation Service (USDA-NRCS) 590 Nutrient Management Practice Standard (USDA 2003). The index is intended to provide stakeholders (including regulatory field staff, watershed planners, and land users) with a tool to assess various landforms and management practices for the potential risk of P transport into surface waters.

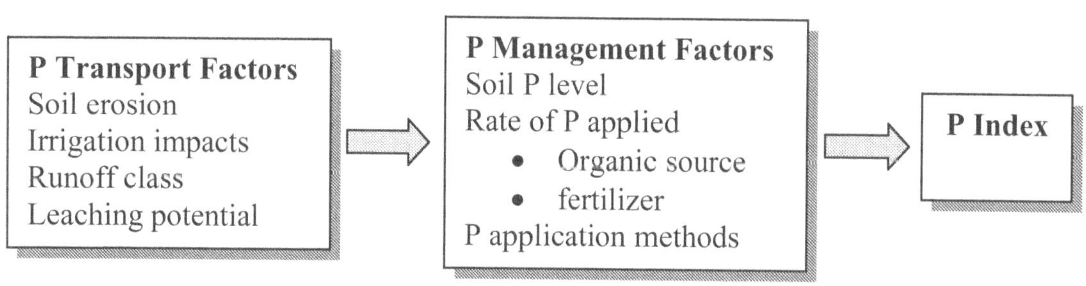

Figure 1.2 P Index concept development

Table 1.2 shows the specific soil and hydrology factors the original P Index considers:

- Soil erosion,
- Irrigation erosion,
- Runoff class,
- Soil P level,
- P fertilizer application rate,
- P fertilizer application method,
- Organic P source application rate and organic P source application method.

Some versions of the P Index adopted by individual states also include factors such as proximity to water bodies and the presence of grass buffer strips.

A P loss rating (weighted value) is then assigned to each factor, depending on unique site conditions. Each factor is then multiplied by the weighted value. Totalizing the weighted factors generally allows the site's potential for P transport from the site to be determined as low, medium, high, or very high as shown in Table 1.3.

Once the site's potential for P transport has been determined, appropriate P management techniques can be applied at higher risk sites. Some of the more common management techniques include conservation tillage, crop residue management, cover crops, buffer strips, contour tillage, runoff water impoundment, and terracing. These strategies are effective for controlling both particulate P and dissolved P (Novak et al. 2002) in runoff. In order to control dissolved P transport to surface water, new best management practices (BMPs) must be developed and implemented.

WTR MANAGEMENT ISSUES

Traditionally, WTR management has been an afterthought for many utilities. WTR is often stored on-site in impoundments or lagoons for many years. Only when the capacity of these storage systems becomes limited and their disposal contemplated do they receive focused attention. At some point, all drinking water treatment facilities must identify and implement methods for managing WTR and removing it from the facility.

There are several common methods of WTR management including direct sewer discharge and removal from the site for disposal or beneficial use. Removal from the site usually requires that WTR be dewatered after being removed from the drinking water process. This makes the handling and transportation to ultimate end use or disposal easier and more cost effective. Where space is available and favorable environmental and zoning conditions allow, non-mechanical methods are sometimes used, such as sand drying beds. For facilities with limited space, mechanical dewatering equipment (such as belt filter presses, centrifuges, plate and frame presses) are utilized for dewatering.

Traditional disposal methods for dewatered WTR such as landfilling may be a concern due to costs or concerns with potential cleanup costs if the landfill is subject to remedial action. Fortunately, recent research has identified unique chemical and physical characteristics of WTR that can lead to use of these materials in manufactured products and the development of successful re-use programs. Cornwell, Mutter, and Vandermeyden (2000) completed an Awwa Research Foundation (AwwaRF) project that identified numerous beneficial use markets and summarized the regulatory climate throughout the U. S.

Table 1.2
P Index matrix

Site Characteristic (weighted value)	Phosphorus Loss Rating (Value)				
	None (0)	Low (1)	Medium (2)	High (4)	Very High (8)
Soil Erosion (1.5)	Not Applicable	<5 Tons/AC	5-10 Tons/AC	10-15 Tons/AC	>15 Tons/AC
Irrigation Erosion (1.5)	Not Applicable	Tailwater Recovery or QS<6 for very erodible soils or QS<10 for other soils	QS>10 for erosion resistant soils	QS>10 for erodible soils	QS>6 for very erodible soils
Runoff Class (0.5)	Negligible	Very Low or Low	Medium	High	Very High
Soil P-Test (1.0)	Not Applicable	Low	Medium	High	Very High
P-Fertilizer Application Rate (0.75)	None Applied	1-30 P_2O_5 Lbs/AC	31-90 P_2O_5 Lbs/AC	91-150 P_2O_5 Lbs/AC	> 150 P_2O_5 Lbs/AC
P-Fertilizer Application Method (0.5)	None Applied	Placed with planter > 2"	Incorporated immediately before crop	Incorporated > 3 months before crop or surface applied < 3 months before crop	Surface applied > 3 months before crop
Organic P Source Application Rate (1.0)	None Applied	1-30 P_2O_5 Lbs/AC	31-60 P_2O_5 Lbs/AC	61-90 P_2O_5 Lbs/AC	> 90 P_2O_5 Lbs/AC
Organic P Source Application Method (1.0)	None	Injected deeper than 2"	Incorporated immediately before crop	Incorporated > 3 months before crop or surface applied < 3 months before crop	Surface applied to pasture or > 3 months before crop

Q = flow rate of water introduced into the furrow (in gallons per minute)
S = furrow slope (in feet per 100 feet, percent).
Example: For a 5 gpm flow rate and a 2% furrow grade: QS = 5 gpm * 2% grade = 10

Source: Adapted from USDA-NRCS 1994

Table 1.3
P Index score relation to site vulnerability for P loss

Total of Weighted Rating Factors	Site Vulnerability
<8	Low
8 – 14	Medium
15 – 32	High
> 32	Very High

Source: Adapted from USDA-NRCS 1994

While considerable advances have been made to understand potential beneficial WTR uses, compliance with some recently promulgated regulations may actually increase the amount of WTR generated by a facility. In particular, facilities practicing Enhanced Coagulation may experience substantial increases in WTR volumes and also change WTR characteristics (U.S. USEPA 1998).

An increased focus on WTR management due to increased quantities, higher disposal costs, and the desire to exploit unique chemical and physical characteristics has led many utilities to consider beneficial use options. One potential beneficial use is direct land application.

The AwwaRF report *Commercial Application and Marketing of Water Plant Residuals* (Cornwell, Mutter, and Vandermeyden 2000) noted that at least 25% of responding utilities reported land application of residuals. The beneficial uses of these applications were listed as improving soil structure, soil pH adjustment, trace mineral addition, improving soil moisture holding capacity, and soil aeration.

All of the utility co-sponsors are currently practicing some form of beneficial use at one or more of their facilities including blending with organic compost to manufacture soils and top dressings, land application for the recovery of strip mined land, cement manufacturing, and direct land application as a soil conditioner. The utilities participating in this study have an especially keen interest in nutrient management.

Many Pennsylvania American Water Company systems lie within the extremely sensitive Chesapeake Bay watershed. The Chesapeake Bay Program (established in 1983) recently updated program goals (Chesapeake Bay Program 2004) to sharply reduce the amount of sediments and nutrients flowing into the Bay. The target for the Bay's P load is a roughly 33% reduction. Pennsylvania is among the compact's state signatories. Other facilities are located in the Ohio River watershed, where similar concerns exist.

The Tulsa Metropolitan Utility Authority (TMUA) is a primary participant in the Watershed Restoration Action Strategy for Eucha/Spavinaw Watershed program. This watershed is a 415-square mile drainage basin that drains into Lakes Eucha and Spavinaw, which are sources of supply for the City of Tulsa. Both lakes have become nutrient enriched leading to increased phytoplankton growth, which in turn has lead to continual taste and odor problems. One possible major contributor of P to the creeks feeding these two lakes is the P-rich manures produced by massive poultry growing operations in the watershed. TMUA and its collaborative partners are actively seeking ways to reduce the introduction of nutrient-laden runoff into the watershed.

Denver Water operates three surface water treatment facilities (Marston, Moffat, and Foothill) that serve the greater Denver metropolitan area. WTRs from these facilities are managed in several ways; some WTRs are blended with soils and/or organic material at a local processing facility. The blended materials are then sold as a commodity to local landscapers.

Denver Water has a unique concern that is common to several Front Range utilities – naturally occurring radioactivity. The presence of radionuclides is an emerging issue for WTR management in Colorado, prompted by the Radionuclides Rule (USEPA 2000) and the release of draft guidance to address the management of WTRs that may contain radionuclides. Radiation exposure issues are of particular concern when WTRs are composted with organic materials. The workers at these facilities may be at risk to radiation exposure due to inhalation (of particular concern with alpha emitting radiation) and point-source exposure to beta and gamma rays.

Denver Water's WTR could be directly land applied or co-applied with organic materials to control the release of P to the environment, thus addressing some of the radioactivity exposure concerns that exist with some forms of composting. The overall impact depends on the soil or background radioactivity, as well as the radioactivity of the WTR. With any beneficial use, the goal would be to show that when the WTR was blended with either material, the impact from the WTR on radioactivity was negligible, or that the activity was below the guideline of 40 pci/L total alpha, as per the existing Colorado Department of Public Health and Environment (CDPHE) WTR beneficial use regulations (CDPHE 2004). This regulation does contain provisions for beneficial use, which require the submittal of a Beneficial Use Plan. The co-application of WTR with biosolids is regulated under Colorado's Biosolids Regulations (CDPHE 2003).

WTR AND PHOSPHORUS CONTROL STRATEGIES

To address this issue and control dissolved P transport to surface water, new control strategies – BMPs - must be developed and implemented. One possible BMP is to remove dissolved P from runoff by applying P-sorbing materials such as aluminum (Al) or iron (Fe) oxide or hydroxide to land. Previous research clearly shows that WTRs have a large P adsorption capacity (DeWolfe 1990, Elliott et al. 1990, Ippolito et al. 2003). Drinking water treatment residuals (WTR) that contain Al or Fe oxide may be beneficially used as a BMP to protect surface water quality by removing dissolved P from agricultural runoff water (DeWolfe and Brandt 1997).

Surface application of WTR has reduced soluble P in runoff water from pasture land that has received manures with poultry manure (Peters and Basta 1996, Basta and Storm 1997, Daniel et al. 1999, Gallimore et al. 1999). The most effective surface application of WTR was as a buffer strip (44.8 Mg WTR/ha) applied to pasture treated with poultry manure (Basta, Dayton and Gallimore; Gallimore et al. 1999). Using WTR, the soluble P was reduced from 15 mg/L to 8.1 mg/L, but large differences (19 and 46 percent) in soluble P reduction were found between the WTR from the two treatment plants used in the study (Gallimore et al. 1999). Drinking water treatment plants (WTPs) use different source water and different treatment chemicals producing WTR with different chemical compositions and P sorption capacities.

Incorporating WTR into soil has reduced excessive soluble P resulting from manure application (Peters and Basta 1996). P solubility in biosolids was reduced by co-application with WTR (Ippolito, Barbarick, and Redente 1999). Application of alum (aluminum sulfate) and ferric sulfate WTR has been shown to eliminate P leaching from biosolids-amended sandy soil (O'Connor and Elliott 2000). Research studies have shown that WTR can potentially serve as a BMP to reduce P Index terms and potential threat of P runoff to surface waters. Incorporation of WTR into soil has been shown to reduce soluble P and soil test P (DeWolfe 1990, Peters and Basta 1996, Morris and Hyde 2001), including in soils without a history of manure applications (Novak and Watts 2004). Co-blending WTR and sewage biosolids reduced soluble P (Ippolito and Barbarick 2001), thereby decreasing an important P source term in the P Index model. Surface application of WTR to pasture land reduced P transport by up to 75% (Gallimore et al. 1999), thus effectively reducing P Index transport terms.

PROJECT OBJECTIVES

As part of a national initiative, NRCS state offices across the United States have been developing state-specific P Indices to improve P management of animal manure and biosolids applications. Some states have incorporated the use of BMPs into the calculation of the P Index. BMPs that reduce P source terms or P transport terms reduce the P Index score and may reduce regulatory restrictions on land application of manure or biosolids in certain situations. One approach proposed by the researchers is to use WTR as a BMP to reduce soluble P runoff from agricultural land and allow for quantitative reductions in P Index scores.

The use of WTR can serve as an effective BMP in states that will be using a P Index to manage agricultural lands receiving large amounts of animal manure from CAFOs. The P Index approach will likely be used in regulatory oversight associated with disposal of animal manure from CAFOs for most states in the near future.

Although the use of WTR has shown great potential as a BMP, the ability of WTR type or land application method to decrease calculated P Index scores has not been reported. However, approaches that use reactive aluminum oxide in soil to estimate soluble P terms in calculation of P Index in Vermont have recently been reported (Jokela 2000). In other recent work, the reactive aluminum content of WTR was strongly correlated with the ability of aluminum based WTR to reduce P runoff from manured soil (Dayton, Basta and Jakober 2001). In this project, estimates of the reduction of P Index associated with WTR type or land application method will be determined.

This AwwaRF Tailored Collaboration (TC) project investigated the beneficial use of WTR to control NPS P runoff and protect surface water quality. Specifically, several types of WTR (aluminum and iron based) were investigated to determine their potential for beneficial use to control dissolved P in runoff-water and reduce P solubility in soil or organic materials.

Various land application methods will be evaluated and calibrated, including incorporation of WTR into the soil, surface application of WTR (broadcast or buffer strip), and co-application of WTR with organic material. Other activities include development of tailored recommendations for WTR use by participating utilities. This report also is intended for general water treatment industry dissemination and use. Specifically, the project was designed to:

- Evaluate different WTR to reduce P loss from agricultural land and lower P Index scores.
- Develop recommendations and land application procedures for beneficial use of WTR based on chemical characteristics of different types of WTR.
- Develop tailored recommendations for beneficial use of WTR as BMPs.

CHAPTER 2
MATERIALS AND METHODS

RESEARCH APPROACH

Research studies utilizing water treatment residuals (WTRs) from the three co-sponsoring utilities were conducted at Oklahoma State University and The Ohio State University (subcontracors), and were coordinated by McGuire Environmental Consultants, Inc. (MEC, project sole contractor). MEC also coordinated activities for processing and shipping of WTR, and collecting and reporting of data to AwwaRF and to each co-sponsor. Research results were intended to provide information on expected performance of different types of WTR to serve as BMPs for soil P control based on chemical characteristics of the WTR and water treatment operational data. As cited in Chapter 1, three co-sponsoring utilities served as participants:

- Pennsylvania-American Water Company (PAWC), primary sponsor
- Tulsa Metropolitan Utility Authority (TMUA), co-sponsor
- Denver Water (DW), co-sponsor

A variety of aluminum- and iron-based WTR were selected to evaluate, and the research was conducted in several sequential phases. These phases are described in the remainder of this section in the order outlined as follows.

- Preliminary WTR characterization
- Plant operational data acquisition
- Extended WTR characterization
- Soil incorporation and organic material blending tests
- Greenhouse studies
- Data interpretation

WTR Selection

The three co-sponsoring utilities each have unique interests with respect to the potential role WTR can have in beneficial use programs directed towards land application for the control of soil P. Both DW and Tulsa MUA employ formal dewatering facilities to maximize WTR solids content prior to removal off-site. PAWC operates nearly two dozen facilities (at the time this research was conducted) and handles both dewatered and liquid WTR in their management program. The DW and TMUA facilities are located near Denver, CO and Tulsa, OK, respectively, while the PAWC facilities are located throughout Pennsylvania. PAWC headquarters are located in Hershey, PA.

Preliminary WTR Characterization

WTR from all participating sponsors was collected and sent to the testing laboratory. The materials were then prepared for subsequent characterization. Eighteen WTRs from PAWC, three WTRs from DW, and two WTRs from TMUA were evaluated. From these WTRs, four PAWC WTRs, two DW WTRs, and the two TMUA WTRs were selected for extended characterization and other subsequent steps. A decision matrix was used to select the four PAWC WTRs. The matrix included such factors as:

- Geographic location within the state
- Location relevant to potential land application sites
- Unique WTR management issues based on past experiences
- Current management costs
- Unique chemical treatment, i.e., use of powdered activated carbon.
- Results of preliminary characterization.

The two DW WTRs were selected based on similar decision factors. Preliminary characterization tests included aluminum and iron oxides via acid ammonium oxalate extraction ((AOE) McKeague and Day, 1966), and determination of Pmax. Both analyses are described under Extended WTR Characterization later in this chapter. Two other important factors impacted the selection of WTRs from PAWC:

- Form of the WTR prior to disposal/reuse – liquid or dewatered
- Major inorganic metal base of the coagulant used in treatment – aluminum or iron.

The form of the WTR was of particular interest since infiltration rates into and hydraulic conductivity within a soil could have an impact on both the rate of reaction and effectiveness of binding soil P. The type of coagulant was of significant interest since the type of metal hydroxides – aluminum or iron – may have different impacts on soil P based on the amount of WTR applied, or the rate of reaction with soil P.

Plant Operational Data Acquisition

A questionnaire was distributed to each co-sponsoring utility requesting specific operational data for the facilities that submitted WTR for preliminary characterization. The questionnaire was distributed electronically with the responses entered into a database. Once completed, queries were executed to correlate the analytical data with plant operational data in an attempt to develop a predictive model for WTR suitability as a BMP for controlling agricultural P runoff. A copy of the questionnaire is provided as Appendix A. The questionnaire requested data in the following categories:

- General plant information
- pH adjustment chemicals
- Raw water characteristics
- Coagulant type and dosages
- Finished water characteristics
- Pretreatment chemicals
- WTR management methods
- WTR regulatory status
- Polymers – type, dose and use - coagulant, flocculent and/or filtration aids

WTR PROCESSING AND PREPARATION

WTR samples from the cosponsors were collected and transmitted using standard chain of custody protocols. Samples were air dried upon delivery to the laboratory as the first step to prepare them for analyses and subsequent research. Dried samples were stored under ambient conditions in an air conditioned room until use in other tests. Further processing varied with WTR analyses and use in research tests. WTR used for soil incubation studies were crushed using a jar mill to pass a 60 mesh sieve (< 250 μm). Air dried WTR for the runoff study was crushed so that all would pass a 0.25 inch mess screen.

Liquid samples were refrigerated upon arrival at 4°C, and mixed to homogenize and applied without sieving. Mixed WTR was subsampled and percent solids determined on subsamples to ensure homogeneity.

EXTENDED WTR CHARACTERIZATION

Once the candidate WTRs were identified for PAWC and DW, extended characterization was conducted on a total of eight WTRs. Data resulting from these analyses were incorporated into the project database and assessed for correlations. The extended characterization included the following analyses:

- Determination of Pmax via linearized Langmuir isotherms
- Total metals via USEPA Method 3051 (USEPA 1994). Metals included:
  - Aluminum
  - Cadmium
  - Iron
  - Calcium
  - Arsenic
  - Copper
  - Phosphorus
  - Zinc
  - Chromium
  - Lead
  - Potassium
- Toxicity characteristic leaching procedure (TCLP) metals, which included:
  - Arsenic
  - Chromium
  - Barium
  - Cadmium
  - Silver
  - Selenium
  - Lead
- Salinity (indicates the amount of soluble salt in soil)
- pH
- Ammonium and nitrate
- Standard soil fertility test (conducted by Agricultural Analytical Services Laboratory of The Pennsylvania State University, described later in this chapter). This is a predictive management tool for agricultural soils that estimates the portion of the crop nutrient requirement that must be supplied as fertilizer for the growing season.

Chemical Methods and Procedures

The following methods and procedures were followed for laboratory research associated with extended characterization and for samples acquired from all soil and organic material treatment methods (described later in this chapter).

Langmuir P Adsorption Maxima (Pmax)

WTR were crushed to < 150 μm. P sorption isotherms were developed by batch equilibration by mixing WTR (1 g) with 10 mL of P solution (KH_2PO_4) and incubating for 6 days. Six P solutions ranging from 0 to 3 g P/L were used. Isotherm data was fit to the linearized Langmuir model and the Pmax was determined.

Total Metals, USEPA Method 3051

Total metal content of the WTRs were determined using EPA Method 3051, a microwave assisted digestion using nitric acid and 30% hydrogen peroxide (USEPA 1994). A representative sample of up to 0.5 g is digested in 10 mL of concentrated nitric acid for 10 min using microwave heating with a suitable laboratory microwave unit. The sample and acid are placed in a fluorocarbon microwave vessel. The vessel is capped and heated in the microwave unit. After cooling, the vessel contents are filtered, centrifuged, or allowed to settle and then diluted to volume and analyzed using inductively coupled plasma atomic emission spectroscopy (ICP-AES).

Toxicity Characteristic Leaching Procedure

Potential WTR contaminants were measured using the Toxicity Characteristic Leaching Procedure (TCLP) EPA method 1311, followed by ICP-AES analysis (USEPA 1992). Samples were immediately preserved by adding nitric acid to lower the pH of the sample to less than 2.0.

Salinity / Electrical Conductivity

Using methods developed by Rhoades (1996), 20 grams of WTR were combined with 20 grams of deionized water. The sample was mixed for 15 minutes to equilibrate. The electrical conductivity (EC) in deciSeimens per meter (dS/m) of the solution was then determined with a calibrated EC Meter.

pH

Using methods developed by Thomas (1996), 20 grams of WTR were combined with 20 grams of deionized water. The sample was mixed for 15 minute to equilibrate. The pH of the solutions was then read with a calibrated pH meter.

Plant Available Ammonium

Ten grams of WTR was combined with 1 M KCl extract and shaken for 1 hour. Samples were filtered and analyzed colorimetrically on a flow injection analyzer using salicylate and hypochlorite chemistry to determine ammonium (Mulvaney 1996).

Plant Available Nitrate

Ten grams of WTR was combined with 1 M KCl extract and shaken for 1 hour. Samples were filtered and analyzed colorimetrically on a flow injection analyzer using cadmium reduction chemistry to determine nitrate (Mulvaney 1996).

Mehlich 3 Extraction

Mehlich 3 extraction was utilized for soil test P and several other soil test parameters. Samples were processed by combining 2 g of sample with 20 mL Mehlich 3 solution (0.2 M glacial acetic acid, 0.25 M ammonium nitrate, 0.015 M ammonium fluoride, 0.013 M nitric acid, and 0.001 M EDTA) followed by a 5 minute equilibration time (Mehlich 1984) followed by ICP-AES analysis.

$CaCl_2$ Extractable Phosphorus

$CaCl_2$ extractable P has been shown to represent the release of P within a soil solution and P that may travel through a soil in soluble form. This weak ionic extractant was developed to buffer the differing ionic strengths of soil solutions. Soluble P was determined by shaking 1 gram WTR with 25 mL of 0.01 M $CaCl_2$ for 2 hours followed by ICP-AES analysis (Kuo 1996).

Soluble Phosphorus

Soluble P was measured in the composite liquid runoff samples acquired in the simulated rainfall tests. Samples were filtered using a 0.45 um pore size filter, then analyzed by ICP-AES analysis (Kuo 1996).

Percent Dry Solids

One hundred grams of sample was filtered with suction applied until the entire liquid fraction of the sample was removed. The remaining solids were then dried at 104 degrees C to determine the weight of dry solids in sample (USEPA 1992).

Reactive Al and Fe Oxide (Al-ox and Fe-ox)

Amorphous, non-crystalline reactive aluminum and iron oxides (Al-ox and Fe-ox, respectively) in WTR were determined by acid ammonium oxalate extraction (AOE) as described by McKeague and Day (1966). With this method, 0.25 g WTR was mixed with 25 mL of acid ammonium oxalate solution for 4 hours, followed by centrifugation for 10 minutes to clear the supernatant. Aluminum and iron in the supernatant were then measured by ICP-AES.

EXTERNAL LABORATORY ANALYSIS

Characterizations of biosolids and poultry manure, as well as supplemental WTR analyses, were conducted by the Agricultural Analytical Services Laboratory of The Pennsylvania State University. Their selection was based on extensive experience with soil fertility and soil amendment analyses, and the emerging use of amendments for nutrient control.

The supplemental WTR analyses evaluated micronutrient levels and standard soil analyses. The results of these analyses are discussed in Chapter 3 – Results and Discussions. The micronutrient analytes included copper, iron, manganese, and zinc. Standard soil analyses included pH, phosphorus, potassium, magnesium, calcium, and cation exchange capacity (CEC). Phosphorus, potassium, magnesium, and calcium concentrations were determined by Mehlich 3 extraction (described earlier in this chapter). The method for pH also is described earlier in this chapter. CEC was determined via summation of potassium, magnesium, and calcium cations.

High Phosphorus Soil

The soil used for the high P soil blending studies and to create the plots used for simulated runoff studies (both described below) was Tonti silt loam (fine-loamy, mixed, active, mesic Typic Fragiudults), acquired from a poultry farm plot near Neosho, MO. This soil, which has routinely received poultry manure applications on an annual basis, was chosen because it is a benchmark soil in an area of high poultry production and consequent P runoff problems. Soil test P was 313 mg/kg (627 lb/acre) as determined by the Mehlich 3 method. The soil was air dried and passed through a 2 mm sieve and homogenized in a cement mixer for 30 min.

Biosolids

Biosolids were obtained from one of the co-sponsor's wastewater treatment facilities. TMUA operates several wastewater treatment plants, and biosolids were obtained from one that processes biosolids anaerobically. This method was chosen because it is a common biosolids processing method and considered to be a likely method utilized by wastewater plants near each of the co-sponsors. Biosolids that are lime stabilized have substantial amounts of calcium and were not desirable, due to the focus on aluminum and iron hydroxide interactions with P, rather than those that may occur with calcium. The biosolids were analyzed using standard assessments methods by the Agricultural Analytical Services Laboratory of The Pennsylvania State University. The biosolids results are discussed in Chapter 3 – Results and Discussions.

Poultry Manure

Poultry manure from the same farm as the soil (described above) was collected and used in both incubation studies with WTR and simulated runoff studies (both described below). Chicken manure was chosen over other animal manures due to relative stability during storage, ease of collection from nearby farms, and the typically high P levels when compared to other manures. The poultry manure was air dried and homogenized in a cement mixer. The poultry manure was analyzed using standard assessments methods by the Agricultural Analytical Services Laboratory of The Pennsylvania State University. The results of these analyses are discussed in Chapter 3 – Results and Discussions.

WTR TREATMENTS

The three methods used to evaluate land application of WTR to control P loss from agricultural land are presented in Table 2.1. Detailed procedures for each application method are listed below the table. The soil incorporation and blending with organic material studies were conducted as laboratory incubations with controlled temperature and moisture conditions, while the surface application / rainfall simulation was conducted in a greenhouse with controlled temperature and precipitation conditions.

Incorporation of WTR into High P Soil

Prior to implementing a full range of incubations with WTRs, a rangefinder test was performed to confirm the appropriate loading rates that would likely achieve the desired reduction in soil P. For this test, the desired reduction involved reducing the initial soil test P from 313 mg/kg (per Mehlich 3 analyses) to between 30 mg/kg and 200 mg/kg. These two values were chosen because, respectively, they are the soil test P values used by Colorado and Pennsylvania as the first step in determining the need to apply the P Index (Sharkoff, Waskom, and Davis 2003; Weld et al. 2003). Another goal of the rangefinder test was to determine if common loading rates could be used for all WTRs.

Full WTR loading range incubations were then performed using a preselected soil (described earlier in this chapter) with high levels of soil test P mixed with all WTR at four loading rates – 1, 2.5, 5 and 10 percent. All soil treatments were replicated three times, and included a control (no WTR applied). The soil/WTR mixtures were incubated at 30° C for four weeks and were watered every other day to maintain moisture. The loading rates used were selected based on the results of the rangefinder tests, and were similar to rates the researchers have used in previous work. In addition, the rates were similar to rates common for agricultural soil amendments. The two liquid WTRs were evaluated as-is (wet) and dried to determine if there would be any difference in the applied form to impact with soil test P or soluble P.

Table 2.1
Generalized approach to study WTR application methods

Application Method	Approach
Soil incorporation	WTR mixed with high soil test P due to high use of fertilizers. Ability of different WTR loadings to reduce soil test P and soluble P was determined.
Surface application with simulated rainfall	WTR surface applied onto soil recently treated with poultry manure. Rainfall was then simulated to generate runoff. Runoff was then collected for analyses. The ability of different WTR to reduce soluble P in runoff water was determined.
Blending with organic material	Poultry manure and biosolids were blended with WTR. The ability of different WTR loadings to reduce soil test P and soluble P was determined.

Samples are collected periodically during incubation at one, two, and three weeks, and chemical parameters measured to monitor the progress of chemical reactions. Soluble P and Mehlich 3 soil test P were determined on air dried samples withdrawn from each of the treatments to determine the ability of WTR to reduce these values.

Surface Application of WTR with Simulated Rainfall

Small plot studies using simulated rainfall were used to evaluate the ability of WTR to reduce runoff P. Figure 2.1 (Miller 1987) shows the major components of the apparatus used for this effort. Small-scale box plots were established under controlled, greenhouse conditions. Each plot (0.5 m wide, 1 m long, and 0.1 m deep) was filled with 75 kg of Tonti silt loam and packed to 10 cm. This meets the approximate field bulk density as reported by the Barry County, Missouri soil survey (1.3 – 1.5 g cm^{-3} moist soil). Each box had fifteen 6.4mm holes for drainage. Landscapers weed cloth was placed in the bottom of the boxes to prevent soil loss. Each box was planted with a vegetative cover of perennial ryegrass (*Lolium perenne* L.), fescue (*Festuca arundinacea*), and Bermuda grass (*Cynodon dactylon*) to simulate a typical hay crop grown in the area. The grasses were allowed to grow for one month and cut once a week. Clippings from the cut grass were not removed allowing the establishment of a thatch layer to simulate actual field conditions. Each box was then placed on a rack to provide a slope of 5%.

Poultry manure was applied at 8.96 Mg ha^{-1} (4 tons $acre^{-1}$) and air dried WTR was applied in a buffer strip to the down-slope of the box plot. The buffer strip occupied 25% of the plot area. WTR was applied to the strip at 0, 5, 10, or 20 Mg ha^{-1}.

Similar to the soil incubation test, a rangefinder test was performed that helped to guide the WTR loading rates used for the full range study. The initial rangefinder rates were based on the P sorption characteristics of the WTRs, and the P levels of both the soil used to manufacture the plots and the poultry manure used for this test.

For the liquid WTR application, poultry manure was added to the surface of the plots at 8.96 Mg ha^{-1} (4 tons $acre^{-1}$) and homogenized liquid WTR was applied to the entire surface of the plots to provide a dry solids WTR application of 0, 2.5, or 5 Mg ha^{-1}. The liquid WTR was applied slowly to prevent runoff. Although liquid did not run off the surface of the plots, liquid did leach through the bottom of the experimental boxes. Therefore, an additional set of control plots were used for the liquid WTR treatments to account for P that may be lost due to leaching of applied water through the box plots. For the liquid WTR controls, water was applied at volumes equal to those applied with the liquid WTR to boxes treated with 8.96 Mg ha^{-1} (4 tons $acre^{-1}$) of chicken manure. Water for the control plots was also applied slowly to prevent runoff. After liquid WTR was applied, the boxes were allowed to rest for 24 hours before the first simulated rainfall.

The liquid WTRs were also applied as dry so that a comparison between liquid and dry application could be made. For this comparison poultry manure 8.96 Mg ha^{-1} (4 tons $acre^{-1}$) was applied over the entire surface of the plot followed by the application of air dried WTR, at 0, 2.5, or 5 Mg ha^{-1}. Following the application of the poultry manure and WTR, the same volume of water was applied to the plots as was applied with the liquid WTR application and liquid controls. The water was applied slowly to prevent runoff. The plots were allowed to rest for 24 hours before the first simulated rainfall.

Simulated rain was supplied at a rate of 7.0 cm h^{-1} by a solenoid-operated variable-intensity rainfall simulator; runoff water was allowed to collect for 30 minutes after it commenced to acquire a composite sample. Four simulated rainfalls were applied to the same treated and control plots to determine if the WTR would continue to reduce runoff P over time. The rainfalls occurred 24 hours after initial application, 2 weeks after application, 4 weeks after application, and 6 weeks after application. Soluble P was then measured in the composite samples.

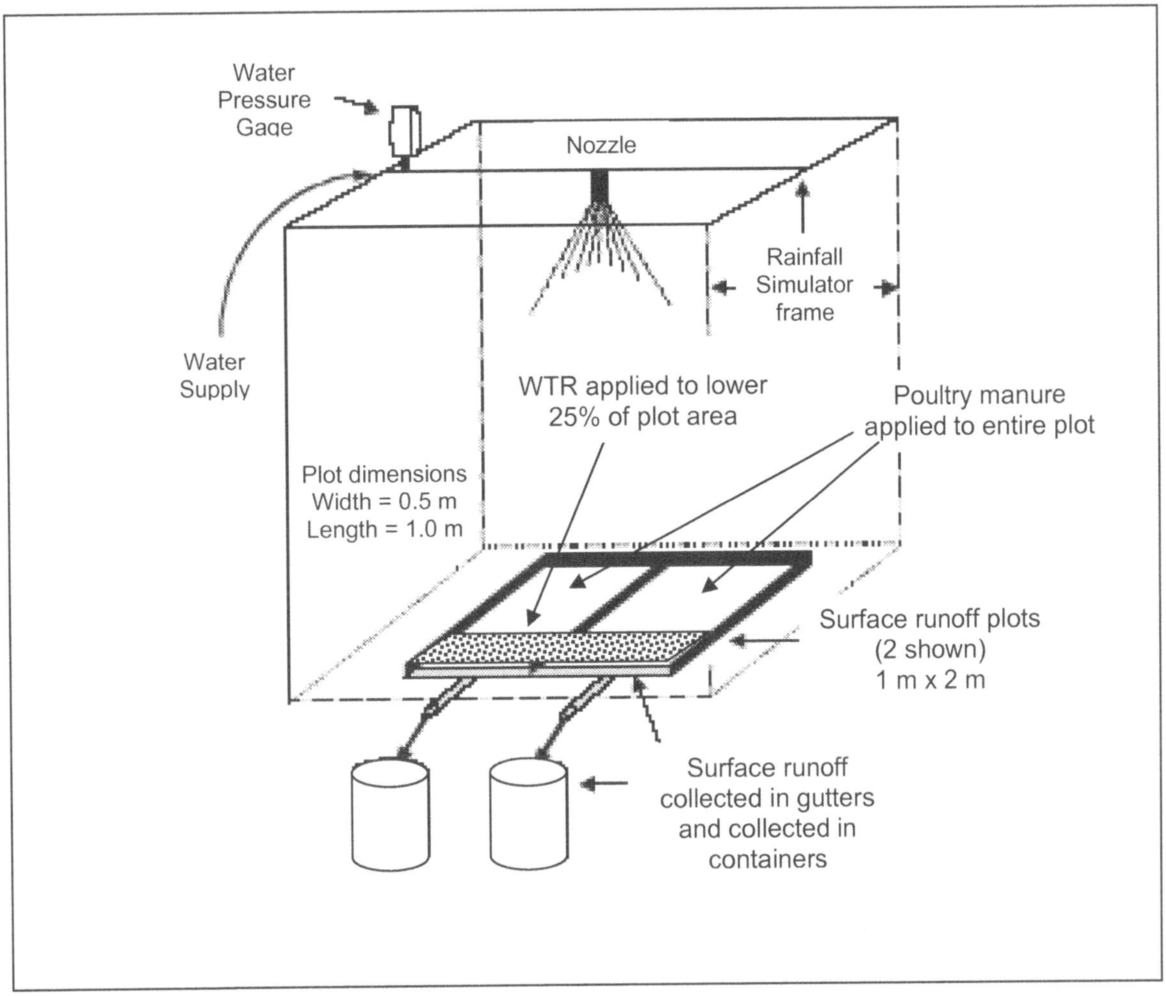

Source: Adapted from Miller 1987

Figure 2.1 Simulated rainfall set-up

Blending WTR with Organic Materials

Biosolids and poultry manure were blended with WTR at several loadings based on P sorption characterization of the WTRs and the P levels of each source of organic material. The loadings used were:

- 5, 12.5, 25, and 37.5% for biosolids incubation tests
- 10, 25, 50, and 75% for poultry manure tests

Blended materials were incubated under controlled laboratory conditions (25° C) for 60 days with a control (no WTR applied). Sampled were collected periodically during incubation, and chemical parameters were measured to monitor the progress of chemical reactions. When chemical reactions were complete (less than 5% difference from last sample), soluble P and soil test P for all WTR treated samples, and the control (biosolid or poultry manure only) were measured.

DATA INTERPRETATION

Data were collected and evaluated throughout the project. Data compiled into the project database were evaluated and correlated to the extent possible to determine interrelations between observed impacts on soil P and WTR characteristics. Also, WTR characteristics and soil P impacts were evaluated with respect to the operating characteristics of the treatment facilities that generated the WTR. The potential impact on the P Index also was determined. Complete results and discussions are found in Chapter 3 – Results and Discussion.

Quality Assurance/Quality Control

The following described the quality assurance and quality control procedures used for this project.

Data Quality Objectives and Criteria for Data

Acceptability of data was determined by following Project Requirements for Precision, Accuracy, Representativeness, Completeness, and Comparability (PARCS) procedures. Specific PARCS parameters were evaluated:

1. Precision was determined by analyzing duplicate samples and performing replicates of experimental procedures. Laboratory duplicates were analyzed at the rate of 10% of the samples. The relative percent difference shall not be greater than 5%.
2. Accuracy was determined by analysis of spiked analytical samples and appropriate reference materials. Spike recoveries were required to be between 80 and 120% and reference materials within 5% of the reported value.
3. Representativeness was evaluated qualitatively.
4. Completeness occurred by ensuring all samples were analyzed and data collected.
5. Comparability was evaluated statistically by Analysis of Variance (ANOVA) techniques with statistical significance level reported.

Sample Handling and Custody Requirements

The purpose of the chain of custody procedure is to ensure reliability and integrity by creating an accurate written record of sample possession from collection through analysis, data reporting, and final deposition. This procedure confirms that samples are collected, transferred, stored, analyzed, and destroyed only by authorized personnel. The chain of custody procedure is important in the event of litigation involving results. It is also useful for routine control of the sample flow. A sample is in custody if it is in any one of the following states:

1. In actual physical possession
2. In view, after being in physical possession
3. In physical possession and locked up
4. In a secure area, restricted to authorized personnel

Sample Collection

The following procedures were used for sample collection:

1. All samples were labeled at the time of collection. The label contained a minimum of the following items: identification number, location, the date taken, the name of the researcher, and any treatment the sample had received.
2. Laboratory notes were used to record (at minimum) the same information used on the sample tag, as well as any field or laboratory measurements, and other information necessary to construct the sample collection process. All sampling time entries were dated and signed, and all pertinent field or laboratory notes were transferred to laboratory files.
3. The sample collector was responsible for the care and custody of the samples until they were transferred to laboratory personnel. The sample collector provided the proper storage conditions and ensured the delivery of all the samples within the permitted holding times. The samples were in physical possession, in view, or stored in a controlled secure area at all times.

Laboratory Custody

All samples received by the laboratory were recorded in a log book denoting the sample description, date received, analysis required, and storage location. The researcher using the collected samples was responsible for sample logging and appropriate storage. In all cases, the laboratory supervisor was responsible for the supervision, care, and custody of a sample once it was relinquished to the laboratory. Upon completing the sample analyses, the unused WTR portion was stored under appropriate conditions until disposition by the laboratory supervisor.

Laboratory Analytical Methods: Quality Control Checks

The quality of laboratory data was assured by a system of internal checks including equipment checks, reagent checks, and laboratory performance checks. These were recorded to verify the operation of the quality control system and to monitor any changes that might have occurred. Subcontracting laboratories were required to adhere to the policies stated or follow their own documented protocol for quality control. The following quality control checks were used.

Matrix Spike and Duplicates

A matrix spike was employed to provide a measure of accuracy for the method used in a given matrix. A matrix spike analysis is performed by adding a predetermined quantity of stock solutions of certain analytes to a sample matrix prior to sample extraction / digestion and analysis. All liquid (water based) samples requiring EPA QA/QC standards included a matrix spike. A matrix spike sample was analyzed with every analytical batch or every 20 samples, whichever was more frequent. A duplicate of the matrix spike sample was analyzed with the matrix spike sample.

Quality Control Charts

Quality control data were kept for routinely analyzed parameters and included standards, precision, and accuracy statements. The mean and standard deviation were calculated. The control data was plotted as quality control charts versus the mean and two and three times the standard deviation.

Blanks

A calibration blank is an aqueous solution that is free of analyte and prepared with the same volume of chemical reagents used in the preparation of the calibration standards and diluted to the appropriate volume used in the preparation of the calibration standard. The calibration blank is used to give the null reading for the instrument response versus concentration calibration curve. One calibration blank was analyzed with each analytical batch or every 20 samples, whichever was greater.

A reagent blank is an aqueous solution that is free of analyte and contains all the reagents in the same volume as used in processing the samples. The reagent blank must be carried through the complete sample preparation procedure and contains the same reagent concentrations in the final solution as in the sample solution used for analysis. The reagent blank is used to correct for possible contamination resulting from the preparation or processing of the sample. One reagent blank was prepared for every analytical batch or for every 20 samples, whichever was greater. For routine soil tests, a reagent blank was used for every analytical batch.

Internal Standards

A standard of known concentration prepared by the analyst was used as an internal standard to monitor and verify instrument performance on a daily basis or with every analytical batch, whichever was more frequent. An internal quality control sample was analyzed with every analytical batch to insure analytical precision.

QC Check Samples

A quality control reference sample is a sample prepared from an independent standard at a concentration other than that used for calibration, but within the calibration range. An independent standard is defined as a standard composed of the analytes of interest from a different source than that used in the preparation of standards for use in the standard curve. A quality control reference sample is intended as an independent check of technique, methodology, and standards and should be run with every analytical batch or every 20 samples, whichever is greater.

To verify digestions, standard reference materials (SRM) were analyzed when the appropriate material exists. The data was documented for accuracy on quality control charts for each parameter. Internal QC samples were also carried through the digestion process. The data was documented as precision on quality control charts for each parameter. Duplicate analyses were performed with every analytical batch or every 20 samples, whichever was greater. The data was documented as "% Recovery" on quality control charts for each parameter.

Calibration Standards

Calibration standards were prepared and used in accordance with approved EPA or other accepted methodology. Standards that did not agree within the limits specified by 10% for other reference materials were not used for analysis. Data from calibration standards were documented on quality control charts labeled "standards" for each parameter. The standard curve was verified by analysis of at least one standard every 20 samples. For matrices other than water, matrix-matched calibration standards were used.

Reagent Checks

Reagents used in sample and standard preparation were verified as free of analyte (below detectable levels) by analysis of the reagent blank.

CHAPTER 3
RESULTS AND DISCUSSION

PRELIMINARY CHARACTERIZATION

Samples of WTR were collected from the utility co-sponsors and processed for subsequent analysis. Analyses for amorphous reactive aluminum and iron oxides (Al-ox and Fe-ox, respectively) via ammonium oxalate extraction (AOE) analyses were conducted to determine the relative P sorption potential of the materials. The results of these analyses are shown in Table 3.1. Relative P sorption potential values of low, medium or high were assigned to the WTRs based on prior work conducted by the researchers and the relative values of this group. High potential for aluminum-based WTRs is generally considered to be greater than 80-90 g/kg Al-ox, while high potential ferric-based WTRs are adjudged high if their Fe-ox levels are over 200 g/kg (see Table 3.1).

The data in the table reveals that all WTRs from Denver Water (DW) possess relatively high P sorption potential, while those from Tulsa Metropolitan Utility Authority (TMUA) and Pennsylvania-American Water Company (PAWC) vary in their P sorption potential.

These data were utilized to help select the eight WTRs that would be employed in subsequent studies (shown shaded in Table 3.1). Four of the eighteen WTRs from PAWC were selected (two aluminum-based, two iron-based), and two of the three WTRs from DW. Other factors used to select candidates for extended characterization from PAWC and DW plants included:

- Geographic location within the state
- Location relevant to potential land application sites
- Unique WTR management issues based on past experiences
- WTR form prior to ultimate end use (liquid or dewatered), and
- Powdered activated carbon (PAC) use for taste and odor (T&O) control

PLANT OPERATIONAL DATA COMPILATION

A questionnaire was developed and distributed to each co-sponsoring utility requesting specific operational data for the facilities generating WTR that were submitted for preliminary characterization. The questionnaire was developed as a form in a database file (Microsoft Access). Once completed questionnaires were returned, they were compiled into the database. This allowed for the execution of queries to correlate the analytical data with plant operational data. The intent of this effort was to determine if a predictive model could be developed for WTR suitability as a BMP for controlling agricultural P runoff. The questionnaire requested data in the following categories:

- General plant information
- pH adjustment chemicals
- Raw water characteristics
- WTR management methods
- Finished water characteristics
- Coagulant type and dosages
- Pretreatment chemicals
- WTR regulatory status
- Polymers – type, dose and use - coagulant, flocculent and/or filtration aids

Table 3.1
Al-ox and Fe-ox levels in WTRs

Aluminum (Al) Based WTR Utility +	Plant Designation	**Al-ox** mg/kg	**Fe-ox** mg/kg	**Projected** P sorption value#	**WTR** Form
DW	CO-1	128	3.59	High	Solid
DW	CO-2	138	5.39	High	Solid
DW	CO-3	155	1.86	High	Solid
TMUA	OK-1	92.6	9.6	High	Solid
TMUA	OK-2	26.2	5.26	Medium	Solid
PAWC	PA-A1	90.2	12.90	High	Solid
PAWC	PA-A2	100.0	13.70	High	Liquid
PAWC	PA-A3	49.8	5.36	Medium	Liquid
PAWC	PA-A4	56.0	9.32	Medium	Liquid
PAWC	PA-A5	63.0	6.70	Medium	Liquid
PAWC	PA-A6	80.4	7.80	Medium	Liquid
PAWC	PA-A7	89.1	8.11	High	Solid
PAWC	PA-A8	45.0	3.07	Medium	Liquid
PAWC	PA-A9	91.1	7.49	High	Liquid
PAWC	PA-A10	46.5	7.50	Medium	Liquid
PAWC	PA-A11	144.0	4.03	High	Liquid
PAWC	PA-A12	168.0	14.00	High	Liquid
Iron (Fe) Based WTR Utility +	**Plant Designation**	**Al-ox** mg/kg	**Fe-ox** mg/kg	**Projected** P sorption value#	**WTR** Form
PAWC	PA-F1	6.28	131	Medium	Solid
PAWC	PA-F2	0.40	248	High	Liquid
PAWC	PA-F3	1.67	232	High	Liquid
PAWC	PA-F4	1.83	110	Medium	Solid
PAWC	PA-F5	2.92	254	High	Solid
PAWC	PA-F6	0.37	315	High	Solid

\+ DW Denver Water
TMWD Tulsa Metropolitan Utility District
PAWC Pennsylvania-American Water Company

* Ammonium oxalate extractable (AOE) aluminum (Al) and iron (Fe)

\# Qualitative assessment of P sorption capacity

An example of the questionnaire is provided as Appendix A. Correlations from the results of the survey and other data gathered through research are provided and discussed later in this chapter.

EXTENDED CHARACTERIZATION

The extended characterization included the following analyses:

- Determination of Pmax using Langmuir isotherms
- Total metals via USEPA Method 3050. Metals included:
  - Aluminum
  - Iron
  - Calcium
  - Potassium
  - Arsenic
  - Cadmium
  - Chromium
  - Copper
  - Lead
  - Zinc
  - Phosphorus
- Toxicity characteristic leaching procedure (TCLP) metals, which included:
  - Arsenic
  - Barium
  - Cadmium
  - Chromium
  - Lead
  - Selenium
  - Silver
- Salinity and pH
- Ammonium and nitrate
- Standard soil fertility test

A summary of the results from the extended characterization analyses are shown in Table 3.2. It is clear that the relative levels of total aluminum and iron correspond with the type of WTR (aluminum- or iron-based). The variability of other metals is likely due to their presence in source waters or contributions from water treatment chemicals (MacPhee, Cornwell and Brown 2002). Metals levels may also be influenced by the pedologic properties of soils within the watershed (Novak and Watts 2004). Based on the TCLP results, none of the WTRs exhibited hazardous waste characteristics for the parameters evaluated, which is consistent with results documented by other researchers.

All aluminum-based WTRs with the exception of PA-A1 had salinities that would be considered non-saline for silty clay loams or clays (Dahnke and Whitney, 1988). PA-A1 and the iron-based WTRs had salinity values that would be considered only slightly saline for soils. These results were expected and reveal that there is little concern to impact the water uptake of plants based on salinity values.

Table 3.2
Summary of extended characterization data

WTR	CO-1	CO-2	OK-1	OK-2	PA-A1	PA-A2	PA-F1	PA-F2
Pmax, g/kg	26.2	32.6	27.1	18.6	29.3	31.7	19.1	19.7
Total Metals, all in mg/kg								
Aluminum	38,273	115,483	88,496	39,146	125,350	89,120	15,815	8,826
Iron	18,860	19,506	22,453	16,573	26,060	16,306	193,433	230,966
Calcium	13,388	5,895	14,603	54,568	5,853	4,540	14,971	31,761
Potassium	2,750	2,811	2,838	1,919	1,219	547	1,407	1,126
Arsenic	6.6	20.9	41.7	29.9	19.3	13.9	BDL	BDL
Cadmium	BDL	BDL	BDL	BDL	BDL	BDL	BDL	BDL
Chromium	136.4	113.3	84.6	92.2	28.8	204.7	89.3	114.4
Copper	37.8	25.8	48.1	93.2	39.2	135.9	83.5	93.4
Lead	BDL	32.3	BDL	5.3	BDL	BDL	12.8	5.86
Zinc	128	276	134	95.9	606	675	550	181
Phosphorus	524	1,965	1,293	2,476	2,993	4,401	1,106	1,372
TCLP Extraction Metals, all in mg/L								
Arsenic	0.774	BDL	BDL	BDL	0.226	BDL	0.146	BDL
Barium	1.25	1.49	1.96	4.61	1.26	1.31	1.16	0.54
Cadmium	BDL	BDL	BDL	BDL	BDL	BDL	BDL	BDL
Chromium	BDL	BDL	BDL	BDL	BDL	BDL	BDL	BDL
Lead	BDL	BDL	BDL	BDL	BDL	BDL	BDL	BDL
Selenium	BDL	BDL	0.07	0.1	0.14	0.16	BDL	BDL
Silver	BDL	BDL	BDL	BDL	BDL	BDL	BDL	BDL
Ammonium Oxalate Extractable (AOE) in mol/kg								
Al	4.74	5.13	3.43	0.97	3.34	3.71	0.23	0.01
Fe	0.06	0.1	0.17	0.09	0.23	0.25	2.34	4.43
Total (Al + Fe)	4.8	5.23	3.6	1.06	3.57	3.96	2.57	4.44
Salinity, dS/m	1.1	0.7	1.2	1.3	2.6	1.7	2.9	3.1
Nitrate, mg/kg	19.05	3.85	50.8	31.85	1.8	136.4	3.15	17.25
Ammonium. mg/kg	31.8	27.8	100.1	40.0	110.4	30.4	33.9	68.1
Standard Soils Analyses Results, all in mg/kg (except pH and CEC)								
pH, Std. units	7.6	7.1	7.3	7.3	7.1	6.9	7.5	7.4
Phosphorus	1	3	8	18	8	16	1	1
Potassium	52	47	206	55	77	142	61	72
Magnesium	54	64	56	123	72	142	534	292
Calcium	5527	1266	3913	15000	1128	1027	6214	10409
Copper	1.5	0.6	3.2	11.2	1.0	5.4	3.6	7.4
Iron	50.8	19.3	57.1	102.8	111.1	106.2	548.1	436.9
Manganese	35.6	79.0	1683.2	872.4	1846.2	2688.8	348.8	585.7
Zinc	1.7	2.6	3.5	3.7	16.6	53.2	39.4	6.2
CEC (meq/100g)	15.6	7.0	16.0	16.2	6.4	6.5	19.6	17.6

The pH values reflect the conditions of coagulation chemistry at each of the water treatment facilities, and are at desirable values (pH between 6 and 8) with respect to soil pH management. In fact, while not measured in these tests, it is likely that these WTRs may have some calcium carbonate equivalence (CCE) and therefore be able to slightly raise soil pH when land applied.

Nutrient levels were highly variable reflecting the unique combination of source waters and the impact of the water treatment process to remove nutrient compounds. OK-1, OK-2 and PA-A2 have nitrate values that could be considered desirable for nitrogen fertilizers, assuming they were immediately incorporated within a soil matrix. The ammonium levels of OK-1 and PA-A1 also could be considered desirable for nitrogen fertilizers. P and potassium (K) levels were below what is considered optimum in soil for these compounds, with the exception of K for OK-1, which is above optimal levels for soils. The Pennsylvania State University Agronomy Guide (2003) recommends optimal P levels of 30 to 50 ppm for all crops, with optimal K concentrations of 100 to 150 ppm for grain crops and 100 to 200 ppm for forage crops.

When dissolved in water, soil nutrients are either positively charged or negatively charged. The cation exchange capacity (CEC) of a soil refers to the amount of positively charged ions a soil can hold. Soils have a slight excess of negative charge sites due to the presence of clay particles and organic material. Thus the higher the clay content and organic material content, the higher the CEC of the soil. Soils with a high CEC will tend to hold onto the positively charged nutrients better than soils with a low CEC. CEC data for the eight WTRs used in this study show a wide range of values. The values are within the typical range (5 to 20 meq/100 g) for medium textured (silt-like) soils (Street, Watson and Pound 2004).

The levels of certain metals – copper, iron, manganese and zinc – were determined to assess their potential as micronutrients. These analyses are not typical soil analyses. Most agricultural soils do not have micronutrient deficiencies. However, the results may indicate which WTRs have the potential to benefit soils that have marginal levels of these constituents. The analyses were performed on Mehlich 3 extractants. The data shows that there are considerable differences on a regional basis. This observation is fully expected, since the soil particles present in local source waters will vary based on regional geology. For instance, both Colorado WTRs were relatively low in all micronutrients compared to other WTRs (Alloway 1990). All micronutrient were within normal ranges for agricultural soils (Alloway 1990).

Figure 3.1 shows total aluminum, iron, and calcium data for all eight WTRs. It is evident from this data that all the aluminum-based WTRs (CO-1, CO-2, OK-1, OK-2, PA-A1, and PA-A2) had higher aluminum than iron values; and as would be expected, the two ferric chloride-based WTRs (PA-F1 and PA-F2) had higher iron than aluminum levels. This clearly illustrates a major assumption made in the conduct of this work: that the metal base for the coagulant has a major influence on the distribution of these major metals. The variation in total calcium levels (4,540 mg/kg to 54,568 mg/kg) indicates differences in source water content and/or the use of pH adjustment chemicals (lime). When compared to other metal levels, total potassium levels of all WTRs were low (547 to 2,838 mg/kg).

Of particular interest is the relationship of Pmax – the maximum amount of P (P) adsorbed per unit amount of WTR (g P / kg WTR) – to the amount of amorphous, non-crystalline reactive aluminum and iron oxides (Al-ox and Fe-ox, respectively) as determined by AOE. Figure 3.2 shows the relationship of Pmax to Al-ox for the aluminum based WTRs, and the associated linear relationship and correlation factor for the data. From this figure and trend data, a relatively positive correlation is observed.

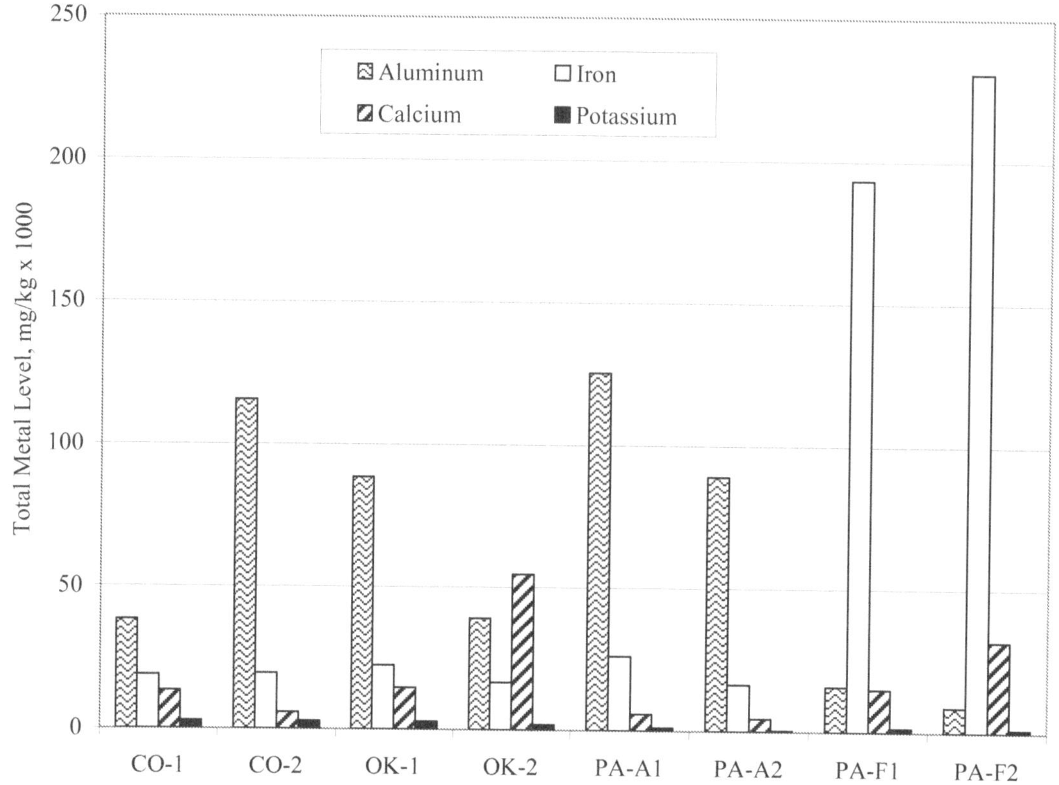

Figure 3.1 Total aluminum, iron, calcium and potassium metals levels

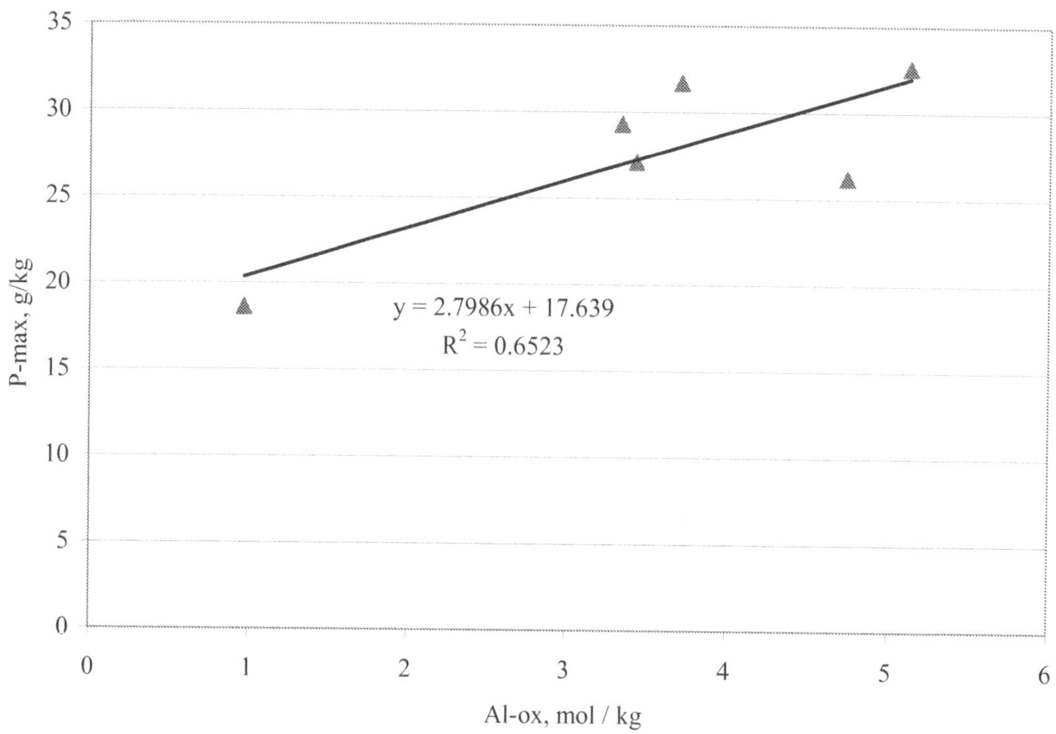

Figure 3.2 Pmax versus Al-ox for six aluminum based WTRs

Figure 3.3 shows a comparison of Pmax to the total Al-ox and Fe-ox for all WTRs (molar basis). The total amount of Al-ox and Fe-ox does not correlate well for the WTR evaluated. No correlations between Pmax and Fe-ox were performed, since there were only two iron-based WTRs used in this research.

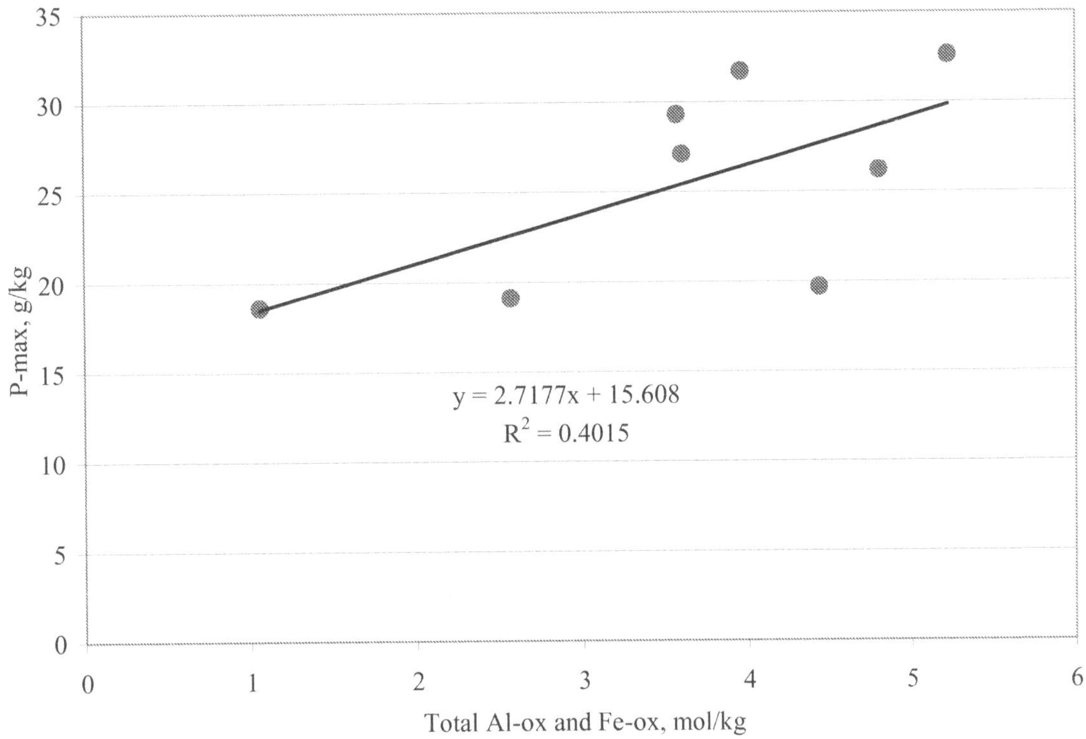

Figure 3.3 Pmax versus combined Al-ox and Fe-ox (molar basis) for all WTRs

The correlation between total aluminum and the reactive Al-ox fractions was evaluated for the aluminum based WTRs and is shown in Figure 3.4. The data indicate that there is no correlation between these two analytes, which is likely due to the use of WTRs that had been processed (dewatered) soon after production. This is significant, since many state regulators base beneficial use via land application on total aluminum, not reactive aluminum. Coupled with the positive correlation between Pmax and reactive aluminum (Figure 3.2), this provides confidence in guiding land application rates of aluminum based WTR on Pmax and reactive aluminum (Al-ox in mol/kg). The correlation between total iron and Fe-ox was not evaluated due to the fact only two iron-based WTRs were evaluated in this research.

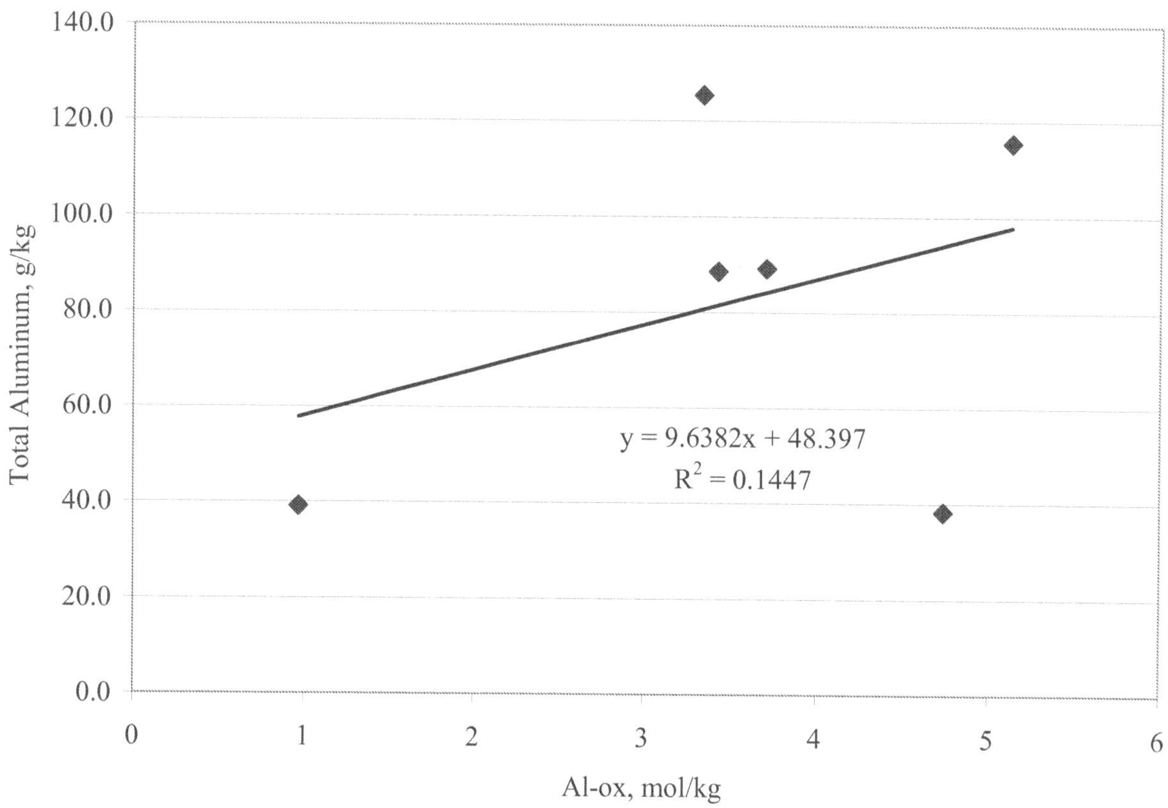

Figure 3.4 Total aluminum versus Al-ox for aluminum based WTRs

WTR TREATMENTS

Incorporation of WTR into High P Soil

Rangefinder Tests

Rangefinder tests were performed to help determine the range of WTR loading rates used for incubation studies that would likely provide a desired reduction of P in a high P soil. This was an important step to determine if common loading rates could be used for all WTRs. The loading rates used for each WTR in the rangefinder test were based on previously determined Pmax values. The loading rates for each WTR and the soil test P (Mehlich 3) and soluble P results for the rangefinder tests are shown in Table 3.3. The results were taken from samples collected four weeks after incubation, and were unchanged from the previous sample (three weeks after incubation), suggesting the reaction between soil P and WTR was complete. It is important to note that both PA-A2 and PA-F2 – the two liquid WTRs - were both dried prior to soil incorporation.

The results show that in general, WTR decreased both soil test (Mehlich 3) P and soluble P. With the exception of CO-2, greater reductions were achieved on soluble P than soil test P. This is expected because the extraction solution for Mehlich 3 P is a much harsher acid than the calcium chloride solution used for the soluble P extraction; thus it would be able to extract the more chemically bound form of P from the soil. The rangefinder test results suggest higher WTR application should be used for subsequent incubations to achieve greater reductions in both soil test P (Mehlich 3) and soluble P. These results are consistent with earlier WTR studies performed by researchers (Peters and Basta 1996) where applications ranging from 10 to 30% WTR by weight were necessary to achieve greater than 50% reductions in soil test P determined by Mehlich 3 P.

It is desirable to utilize WTR to reduce soil P levels to the point where the P pollution potential is reduced, and soil test P levels are reduced to between 30 and 200 mg/kg (Sharkoff, Waskom, and Davis 2003; Weld et al. 2003). The results suggested that WTR loading rates of 10, 25, 50, and 100 g/kg (1, 2.5, 5 and 10%, respectively) would be appropriate for each WTR (see Chapter 4 for details).

Table 3.3
Incubation rangefinder test results - reduction in soil test and soluble P

WTR	% WTR Added	Mehlich 3 P (mg/kg) *	% Reduction Mehlich 3	Soluble P (mg/kg) *	% reduction Soluble P
None - Control	0	322	0.0	16.8	0.0
OK-1	1.13	294	8.7	10.1	40.1
OK-2	1.72	323	-0.3	16.1	4.5
CO-1	1.19	284	11.9	11.7	30.5
CO-2	0.96	274	15.2	15.7	6.7
PA-A1	1.07	272	15.6	10.9	35.3
PA-A2	4.23	289	10.4	12.8	23.8
PA-F1	1.64	305	5.6	12.6	25.3
PA-F2	2.05	298	7.5	15.0	10.8

* mean values based on all replications after four weeks of incubation

Full Range Incubations

The next phase of soil incubations evaluated the full range of loadings that were established during the rangefinder test described above. The WTR loading rates for each WTR were 10, 25, 50, and 100 g/kg (1, 2.5, 5, and 10%, respectively). During this test, it is important to note that the liquid WTRs – PA-A2 and PA-F2 – were evaluated both in dry form and "as-is" liquid form (approximately 4% solids). For the loading that involved liquid WTRs, deionized water additions were substantially reduced to achieve saturated soil conditions. The results of solid WTRs – those that are dewatered prior to their ultimate fate – are discussed separately from the liquid WTRs. The following observations can be made from the data shown in Table 3.4:

- The results of the incubation tests revealed several important trends and observations. A summary of the impacts of all solid (dewatered) WTRs on both soil P and soluble P over time is shown in Table 3.3. All WTRs had varying degrees of impact on both soil test P and soluble P. This helps to defend the general thesis that each WTR possesses unique P adsorption characteristics.
- All WTRs were able to reduce soil test P to less than 200 mg/kg at the 10% loading rate, and in most cases were able to do this after just one week of incubation time.
- The majority of the soil test P reduction occurred after only one week of incubation. This is likely due to the large surface area available on the solids surfaces of both the WTRs and the high P soil.
- Of the WTRs tested, the most reactive was PA-F1, an iron coagulant-based WTR. It is important that assumptions should not be made about the relative sorption capacity of iron coagulant-based coagulants compared to aluminum-based coagulants, since only one iron coagulant-based WTR was included in the full range incubation.
- The ratio of soluble P to soil test P ($CaCl_2$:Mehlich 3) shows the potential of the soil test P to solubilize and release to runoff water. In general, a decrease in this ratio should be viewed as beneficial and support the use of this WTR; however, the actual value of soluble P and soil test P also should be considered when guiding the rate of application of WTR.

Table 3.4
Summary of the impact of solid (dewatered) WTRs on P

Solid WTR	% WTR Added	Soil Test P (Mehlich 3), mg/kg			Soluble P ($CaCl_2$), mg/kg			$CaCl_2$:Mehlich 3 x 100		
		Weeks of Incubation			Weeks of Incubation			Weeks of Incubation		
		1	2	4	1	2	4	1	2	4
Control	0	315	-	-	13.37	-	-	4.24		
OK-1	1	304.0	305	274	6.24	4.13	2.97	2.053	1.354	1.084
	2.5	297.3	264	232	3.48	2.11	1.55	1.171	0.799	0.668
	5	221.3	191	175	2.43	1.29	0.75	1.098	0.675	0.429
	10	146.1	125	106	1.61	1.03	0.38	1.102	0.824	0.358
OK-2	1	302	327	302	7.95	7.73	7.20	2.632	2.364	2.384
	2.5	310	306	293	5.32	5.55	5.23	1.716	1.814	1.785
	5	253	255	248	3.60	4.17	3.53	1.423	1.635	1.423
	10	203	194	185	2.10	2.77	2.15	1.034	1.428	1.162
CO-1	1	273	286	253	6.58	6.92	4.69	2.410	2.420	1.854
	2.5	304	259	229	4.42	3.87	2.33	1.454	1.494	1.017
	5	257	234	212	1.98	2.53	1.14	0.770	1.081	0.538
	10	200	185	163	1.53	1.63	0.60	0.765	0.881	0.368
CO-2	1	291	308	285	2.39	2.87	2.19	0.821	0.932	0.768
	2.5	268	240	225	0.92	1.53	0.53	0.343	0.638	0.236
	5	159	137	140	0.64	1.29	0.33	0.403	0.942	0.236
	10	81.7	40.0	41.9	0.61	1.33	0.19	0.747	3.325	0.453
PA-A1	1	311	325	303	1.79	2.76	1.80	0.576	0.849	0.594
	2.5	254	250	234	0.83	1.67	0.73	0.327	0.668	0.312
	5	170	124	134	0.70	1.25	0.39	0.412	1.008	0.291
	10	54.3	32.0	42.5	0.68	1.41	1.03	1.252	4.406	2.424
PA-F1	1	267	217	231	5.62	5.23	4.29	2.105	2.410	1.857
	2.5	140	144	141	2.25	2.59	1.71	1.607	1.799	1.213
	5	52.8	54.0	58.5	1.12	1.88	0.88	2.121	3.481	1.504
	10	10.9	4.67	9.86	0.72	1.41	0.63	6.606	30.193	6.508

Tables 3.5 and 3.6 show the impacts over time of liquid WTRs (PA-A2 and PA-F2) on both soil test and soluble P, respectively. The liquid WTRs were evaluated both in a dried form and a wet (as-is, nominal 4% solids) form to determine if there would be any difference in the ability of the WTR to react with P.

The results demonstrate that the wet (as-is) form of WTR had a substantially greater impact on P levels than the dry material at all loading rates greater than 1%. This difference was especially notable for the aluminum coagulant-based PA-A2 at the 5 and 10% loadings, and is likely due to the more amorphous and reactive nature of the aluminum hydroxides (less crystalline and mineralized) when they are in the liquid form. There is also the likelihood that the wet materials have a greater surface area than the dried, irrespective of the fact they were processed to pass a 60 mesh (250 um) sieve. This is a significant observation, since each WTR is currently managed in the liquid form at their facility of origin. While drying these materials would probably lower transportation costs from the water treatment facility to the point of ultimate use or disposal, their value as materials to control soil P would be substantially reduced.

As noted with the solid WTRs discussed above and summarized in Table 3.3, the majority of the impact on both soil test P and soluble P occurred in the first week of incubation. This was particularly evident in the wet form. Again, this observation is likely due to the solids surface areas of both the WTR and soil. Field applications of WTR onto the soil surface would likely have a reduced impact, while incorporation of the WTR with the soil after application would likely aid the reaction between WTR and soil P.

Perhaps the most noteworthy results of the incubation tests are presented in Figures 3.5 and 3.6 that show the correlations between Pmax and percent reductions in soil test P and soluble P, respectively after four weeks for all the solid (six) WTRs used in this study. Data for the liquid WTRs that were dried are not included since their processing would not allow for appropriate comparison with the materials that are dried at the water treatment facilities. The figures illustrate the strong correlation between Pmax and reductions in soil test P at loadings greater than 1% (Figure 3.5), and that considerable correlations existed at all loadings between Pmax and reductions in soluble P (Figure 3.6). This confirms the value of determining Pmax as an indicator of the relative potential for WTR to have an impact on both forms of P.

One other important observations is that as WTR loadings increased, the correlation factor (r-squared value) and slope both increased for the Pmax and soil test P correlation, while the intercept decreased. This suggests that as WTR loading increases (and thus comprises a larger portion of the soil volume), the incremental impact of WTR loading on soil test P will be greater. The impacts on soluble P correlations indicate that as WTR loadings increased, correlations factors were similar while slopes decrease and intercept values increased. This shows that as WTR loadings increased, the observed percent reduction in soluble P will become higher irrespective of the Pmax value.

Table 3.5
Impacts of liquid WTR on soil test (Mehlich 3) P in dried and wet (as-is) forms

Liquid WTR	Loading Rate, %	Soil test P, mg/kg – Weeks of Incubation								
		1			2			4		
		Dry	Wet	% Diff	Dry	Wet	% Diff	Dry	Wet	% Diff
PA-A2	1	285	306	7%	274	281	3%	257	261	2%
	2.5	250	178	-40%	241	174	-39%	235	163	-44%
	5	205	68.0	-201%	189	62.7	-201%	190	64.1	-196%
	10	142	15.9	-791%	108	18.0	-500%	96.7	15.4	-528%
PA-F2	1	308	314	2%	273	289	5%	250	276	9%
	2.5	243	163	-49%	237	154	-54%	227	156	-45%
	5	169	91.1	-86%	177	103	-73%	168	98.6	-70%
	10	97.7	59.3	-65%	102	51.3	-99%	97.3	49.9	-95%

Soil only control (0% loading) soil test P = 315 mg/kg

Table 3.6
Impacts of liquid WTR on soluble P test ($CaCl_2$) P in dried and wet (as-is) forms

Liquid WTR	Loading Rate, %	Soluble P, mg/kg - Weeks of Incubation								
		1			2			4		
		Dry	Wet	% Diff	Dry	Wet	% Diff	Dry	Wet	% Diff
PA-A2	1	6.21	1.30	-378%	4.93	1.03	-379%	3.40	0.95	-260%
	2.5	2.82	0.45	-533%	2.41	0.71	-239%	1.38	0.59	-133%
	5	2.36	0.38	-522%	1.03	0.80	-29%	0.72	0.67	-7%
	10	1.98	0.28	-617%	0.64	0.49	-31%	0.43	0.46	6%
PA-F2	1	7.62	9.07	16%	8.57	6.57	-30%	6.76	5.88	-15%
	2.5	4.68	1.79	-161%	4.99	1.35	-270%	4.04	1.20	-236%
	5	2.35	0.77	-204%	3.12	1.08	-189%	2.31	0.88	-164%
	10	1.14	0.74	-55%	2.45	0.63	-289%	1.35	0.56	-143%

Soil only control (0% loading) soluble P = 13.37 mg/kg

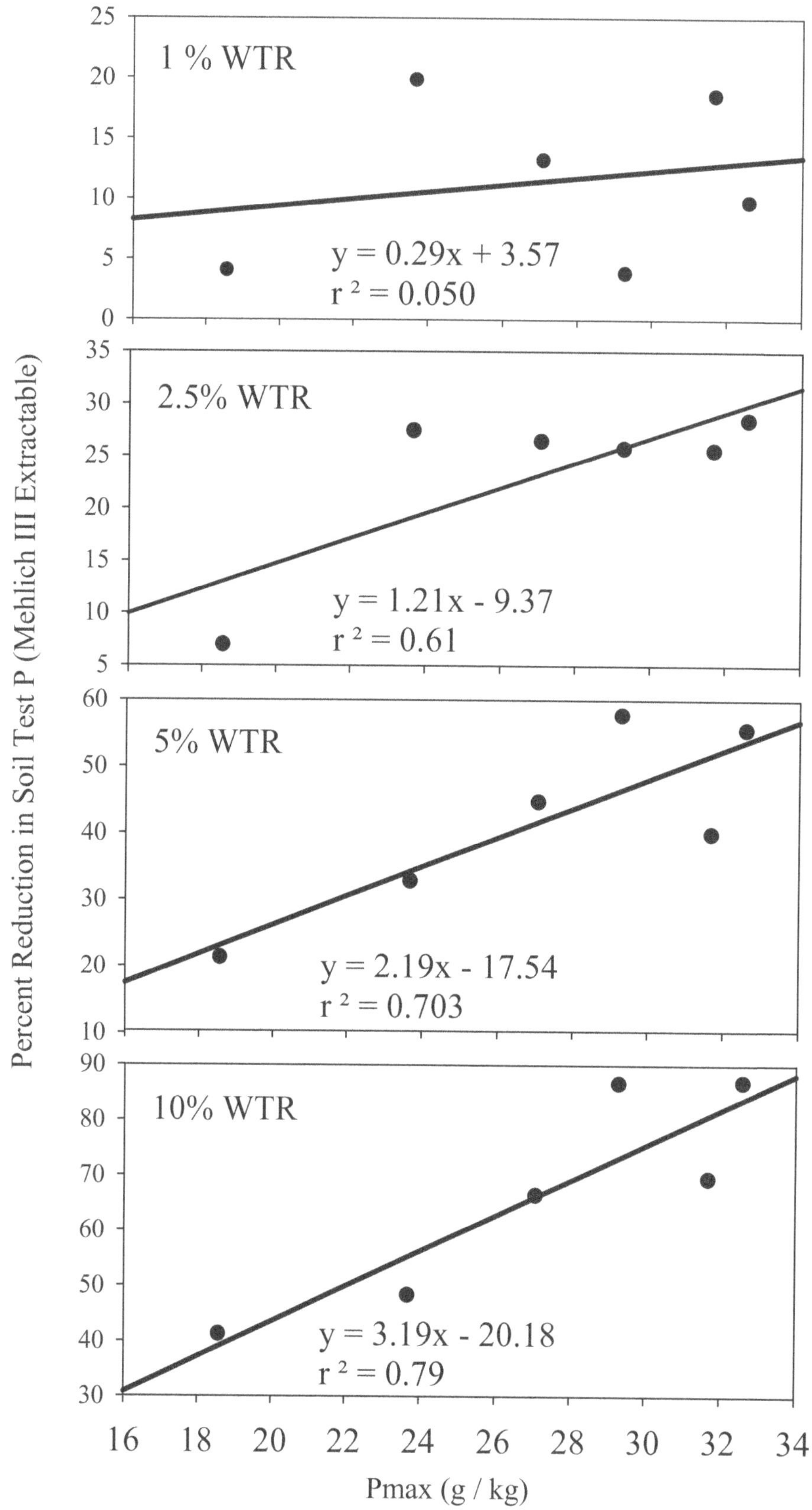

Figure 3.5 WTR blended with high P soil - Pmax versus soil test (Mehlich 3) P reduction

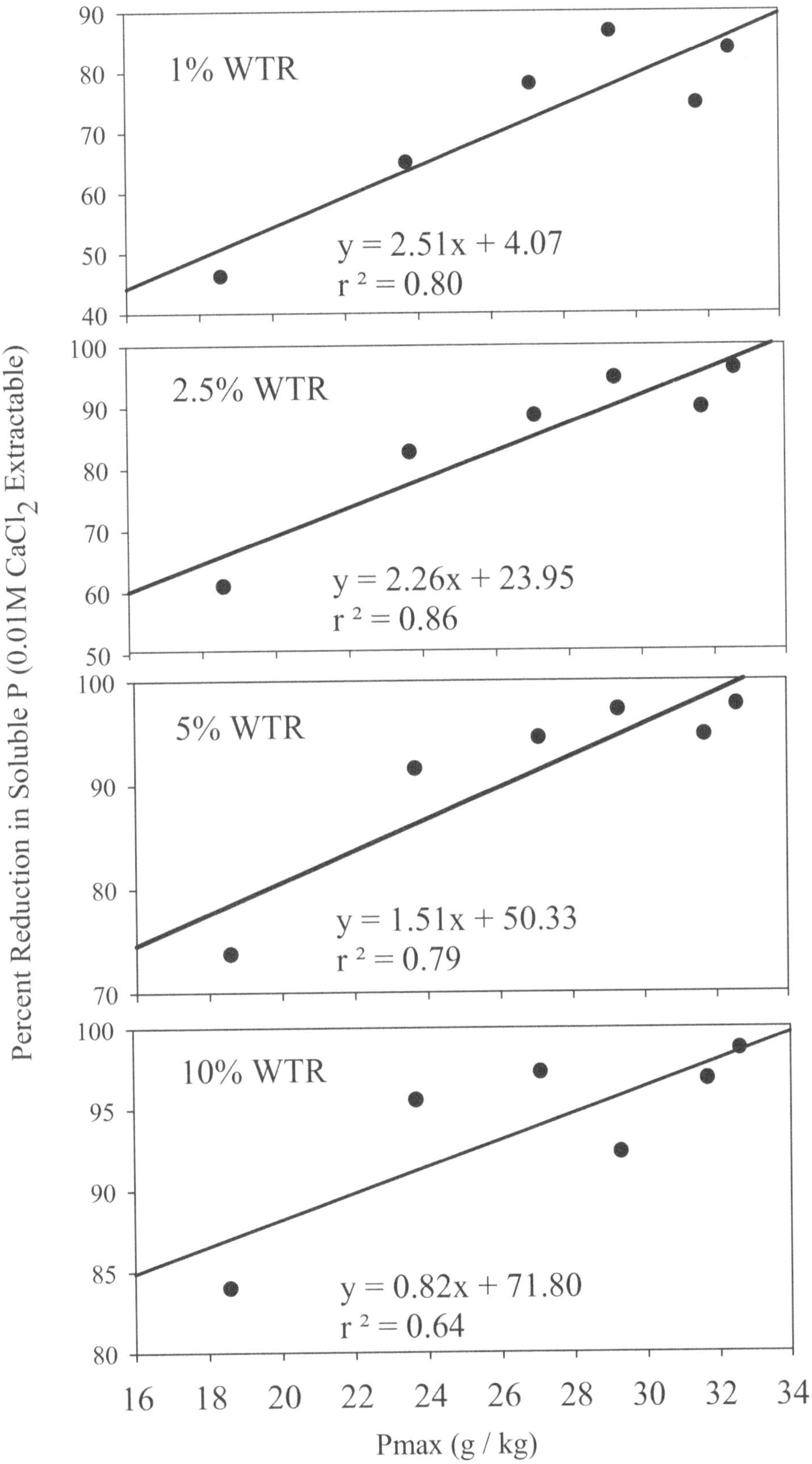

Figure 3.6 WTR blended with high P soil - Pmax versus soluble P ($CaCl_2$) reduction

Surface Application of WTR with Simulated Rainfall

Rangefinder Simulated Rainfall Tests

Similar to the incubation studies, rangefinder tests were performed to help determine the range of WTR loading rates used for surface applications studies that would likely provide a desired reduction of P in a high P soil that also had been treated with a poultry manure containing high P levels. Again, this was an important step to determine if common loading rates could be used for all WTRs.

The loading rate used for each WTR (10 Mg/ha for solid WTRs, 5 Mg/ha for liquid WTRs) in the rangefinder test was based on previously determined Pmax values and a rate that would likely create an impact on soluble P in the runoff water. The soluble P results for the surface application rangefinder tests are shown in Table 3.6, along with the percent reduction accomplished compared to the control. For this test, only one rainfall event was performed.

As seen in Table 3.7, all WTR reduced runoff P. Application of the six solid WTRs at 10 Mg/ha decreased runoff P from 53.0 to 69.0%. Based on these results, future application rates of 5, 10, and 20 Mg/ha were selected.

Large volumes of liquid WTR had to be added to box plots treated with manure to deliver 5 Mg/Ha (solids loading) of WTR. Liquid controls were used with an additional set of control plots to account for P that may be lost due to leaching of water through the box plots. As Table 3.6 shows, this resulted in much lower control runoff P for liquid WTR applications (14.6 mg/L vs. 31.3 mg/L). Application of liquid WTR at 5 Mg/ha (solids loading) also reduced runoff P (81.5% and 28.9 % for aluminum and iron based WTRs, respectively. Application of larger amounts (>5 Mg/ha) would require large volumes of water and is likely impractical due to restrictions on hydraulic loading rates imposed by regulatory permit conditions. However, lower application rates may be practical. Therefore, liquid WTR was applied to deliver solids at 2.5 Mg/ha and 5 Mg/ha in the full range runoff studies.

Table 3.7
Simulated rainfall rangefinder test results - reduction in soluble P

Treatment	Runoff P, mg/L *	% Reduction in Runoff P
Solid WTR, 10 Mg/ha		
Control	31.3	0.0
OK-1	14.7	53.0
OK-2	14.4	54.0
CO-1	9.7	69.0
CO-2	11.6	62.9
PA-A1	10.8	65.5
PA-F1	13.6	56.5
Liquid WTR, 5 Mg/ha		
Control PA-A2	14.6	0.0
PA-A2	2.7	81.5
Control PA-F2	4.5	0.0
PA-F2	3.2	28.9
* mean values based on all replications		

Full Range Simulated Rainfall Tests

The results of the full range simulated rainfall tests are shown in Figure 3.7 for the control (no WTR loading) and three application rates – 5, 10, and 20 Mg/ha. This figure shows that higher WTR application rates resulted in decreased runoff P for all WTR and rainfall times. Rainfall events alone (control) resulted in substantial reductions in runoff P; however, runoff P from the control plot was still substantial after four simulated rainfalls – greater than 3.0 mg/L. The amounts are considered substantial, since soluble P concentrations that can cause eutrophication of surface waters are as low as 0.01 to 0.03 mg/L (Sharpley et al. 1996). The times shown correspond to the time after application of poultry manure as follows:

- 1 – 24 hours
- 2 – 2 weeks
- 3 – 4 weeks
- 4 – 6 weeks

The two liquid WTR (PA-A2 and PA-F2), were applied both in liquid (as-is) and dried forms to determine if there was a difference in P sorption capacity due to physical form. The hypothesis was that liquid WTR would have a greater P binding efficiency due to higher amorphous (hence more reactive) metal oxide concentrations, greater surface area, and homogeneity throughout the liquid that is applied. To match the hydraulic loading due to differences in percent solids, a liquid control was run for each material at each application level. The liquid applications were based on actual hydraulic loading rates currently used for the PAWC WTRs based on regulatory requirements (e.g., there is a PA DEP maximum application rate of 5,000 gallons/acre/day to reduce the potential for liquid runoff).

The results of the liquid vs. dry comparison are presented in Figure 3.8. The control for the dry applications (31.4 mg P/L) was included to illustrate that the additional hydraulic loading of the liquid applications reduced runoff water P considerably before any WTR application. The addition of liquid WTRs caused a further decrease in runoff P. The WTR applied as liquid reduced P slightly better than when dried. This is likely because liquid WTRs have a greater reactive oxide surface area than when dried. This is consistent with observation made in the incubation tests.

As with the incubations results, significant correlations exist between the Pmax values determined during the initial characterization and the results of the simulated runoff tests for the six solid WTRs. The liquid WTRs were not included in developing these correlations due to the difference in their physical form. The correlations between Pmax and the recent reduction for each loading rate used in the simulated runoff study for the six solid WTRs are shown in Figure 3.9. What is particularly revealing is that the slope of the correlation for each loading appear similar (ranging from 1.36 to 1.82); however, this was not statistically evaluated. The figure also provides another perspective on the fact that higher WTR loadings have a greater impact (higher percent reduction) on runoff P.

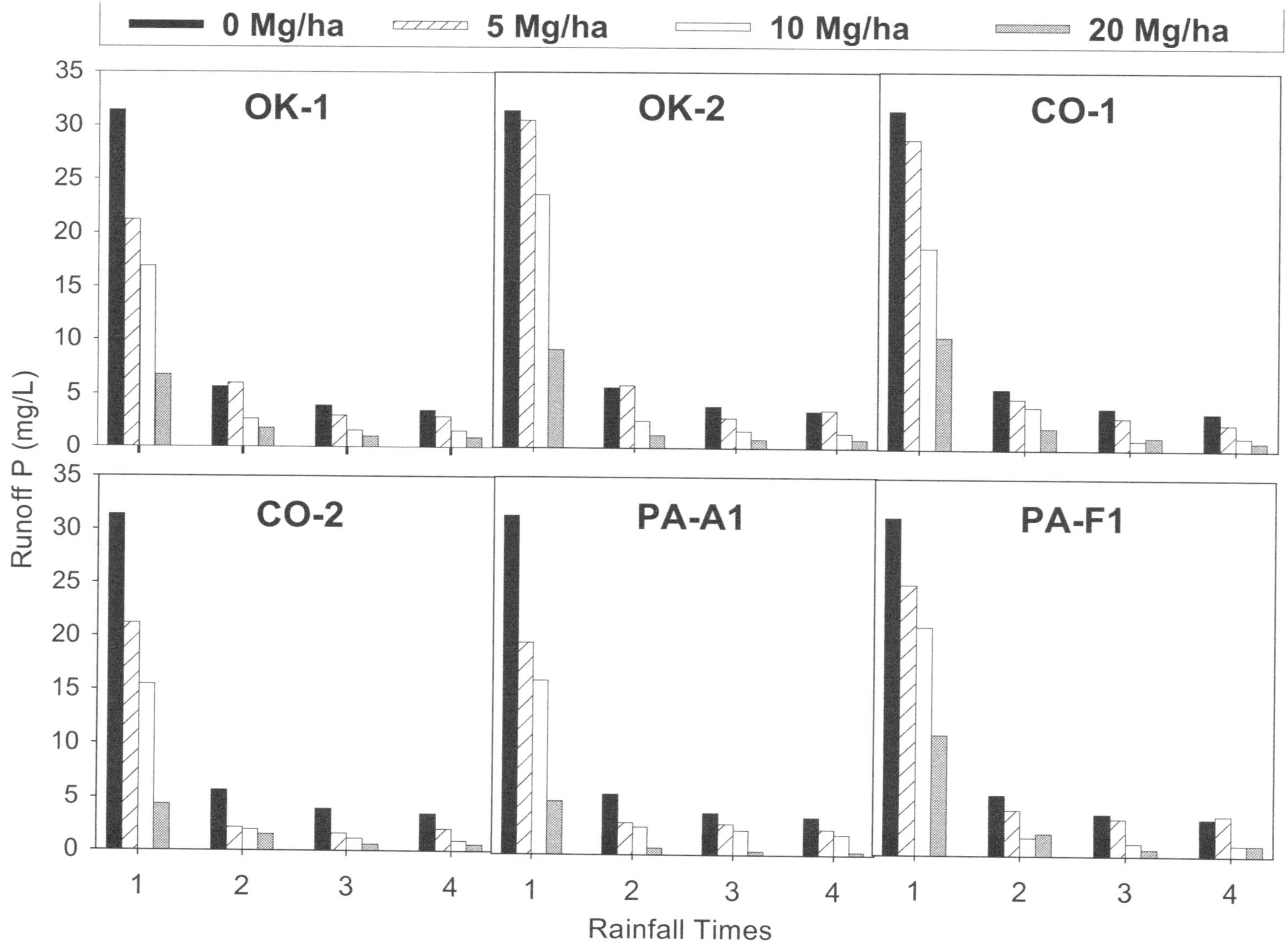

Figure 3.7 Runoff P from plots treated with six solid WTRs at four times after application

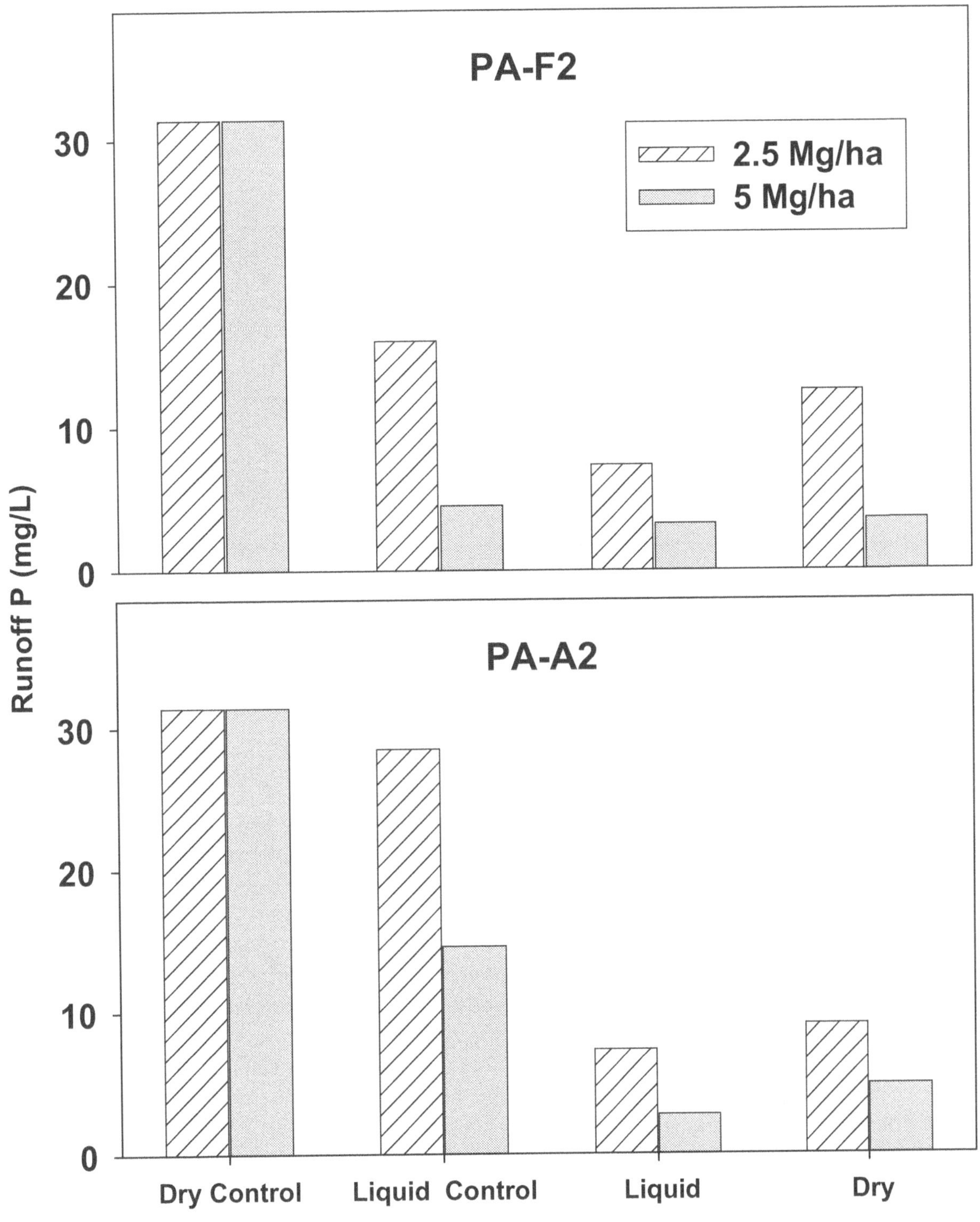

Figure 3.8 Comparison of runoff P reductions by liquid WTR – as-is and after drying

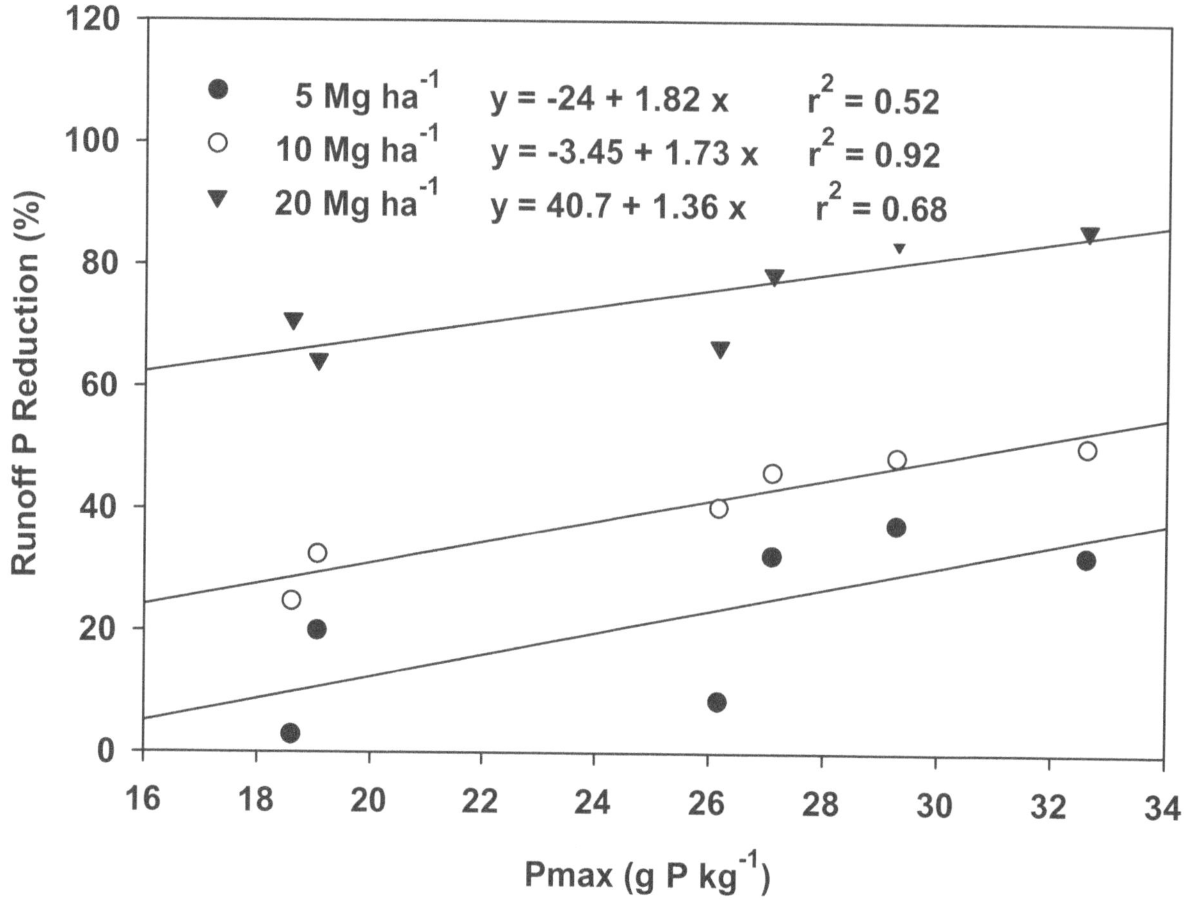

Figure 3.9 Simulated runoff study - reduction of P vs. WTR Pmax for six solid WTRs

Blending WTR with Organic Materials

This portion of the project evaluated blending WTR with biosolids and poultry manure as a possible Best Management Practice (BMP) for managing P in these materials. The studies reveal the potential of WTR to reduce P when they are incorporated into the organic materials. Co-application, where the WTR and organic materials are simultaneously applied, was evaluated in the simulated runoff studies where poultry manure was applied prior to the application of WTR. For these tests, all WTRs were evaluated in a dried form, rather than separately evaluating solid and liquid WTRs, or liquid WTRs in dried and as-is forms. The reason for this was that very high loading rates were used in an effort to determine impacts on both soil test and soluble P. Therefore, liquid applications would not be practical. The data in this study were obtained from analyzing samples acquired 60 days after blending.

A summary of P values for both materials is provided in Table 3.8. The P values for the high P soil used in the incubation tests are shown for comparison; it is clear that the organic materials have substantially higher values of both soil test and soluble P. These extremely high P values illustrated the value in identifying BMP methods to control the release of P from these materials to the environment. Complete characterization information for both the poultry manure and biosolids used for this study are provided in Appendix B.

Table 3.8
Summary of P data for organic materials and high P soil

Material	Soil Test P (Mehlich 3), mg/kg	Soluble P (CaCl2), mg/kg
High P Soil	315	13.4
Biosolid	1,676	62.5
Poultry Manure	10,160	2,053.5

The impact on soluble P and soil test P from blending poultry manure with the CO and OK WTRs are shown in Figure 3.10, while similar results of blending poultry manure with all PA WTRs are shown Figure 3.11.

In general, soluble P levels were substantially reduced by all WTRs at loading rates at or above 25%. At 10% loadings, all PA WTRs were also able to effect a substantial (greater than 70%) reduction in soluble P. Also, most WTRs exhibited little or no incremental increased impact on soluble P above a 25% loading.

The impact of each WTR on soil test P exhibited similar trends to the impact on soluble P. This suggests that the relative ability to impact either form of P was similar for each WTR. The percent soil test P reduction achieved at lower loadings rates (10 and 25%) were generally less that the impact these loading rates had on soluble P. This observation is logical, in that the soil test P extraction is a more rigorous extraction that releases P that is more strongly bound to solids.

The impact on soluble P and soil test P from blending biosolids with the CO and OK WTRs are shown in Figure 3.12, while similar results of blending biosolids with all PA WTRs are shown Figure 3.13. It should be noted that the WTR loadings used for the biosolids were less than those used for the poultry manure test, due to the orders of magnitude difference in P levels between the two substances as shown in Table 3.8.

As with the poultry manure tests, the impact on P levels (percent reduction) increased as WTR loadings increased. This trend was similar for both soil test P and soluble P. This again indicates that the relative ability to impact either form of P was similar for each WTR. What is particularly interesting about these data is that the lowest WTR loading rate of 5% resulted in at least a 40% reduction in soil test P for each WTR evaluated in this study, while also achieving at least a 30% reduction in soluble P. The ability to impact soluble P is of particular importance, since many soil nutrient management agencies are considering limiting application rates of organic materials based on soluble P levels.

Practically, it is likely that most drinking water utilities would generate WTR at rates that could match the lower loading rates used for both the poultry manure and biosolids tests. The correlations between Pmax and the percent reductions at these lower loading rates (5% for biosolids, 10% for poultry manure) were essentially nonexistent. This is more than likely due to the complex chemical nature of biosolids and poultry manure, and factors associated with the biostability of organic materials. Nonetheless, the data does demonstrate that WTR have the potential to reduce both soil test P and soluble P of these materials.

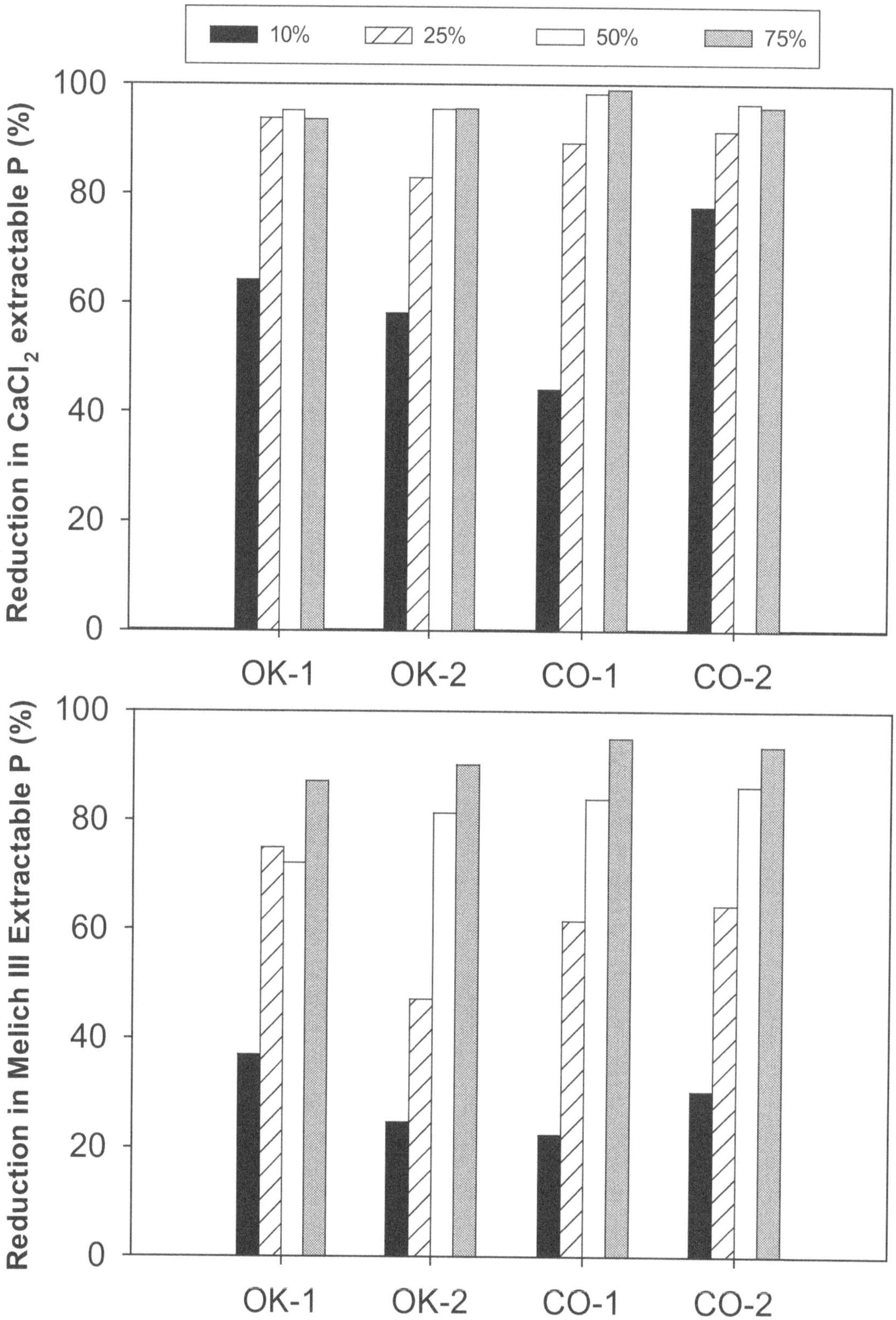

Figure 3.10 WTR blended with poultry manure – reduction in P for OK and CO WTRs

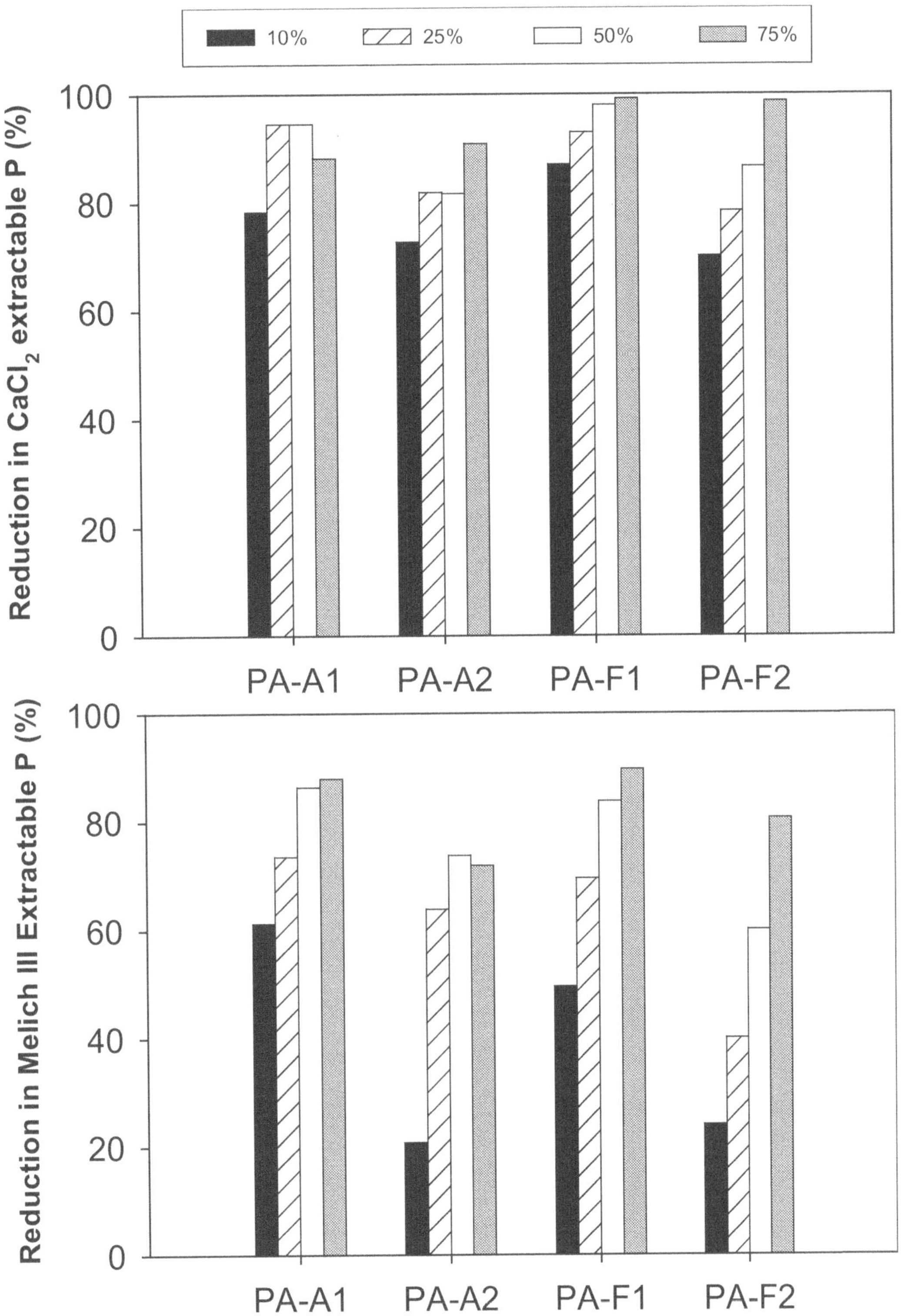

Figure 3.11 WTR blended with poultry manure – reduction in P for PA WTRs

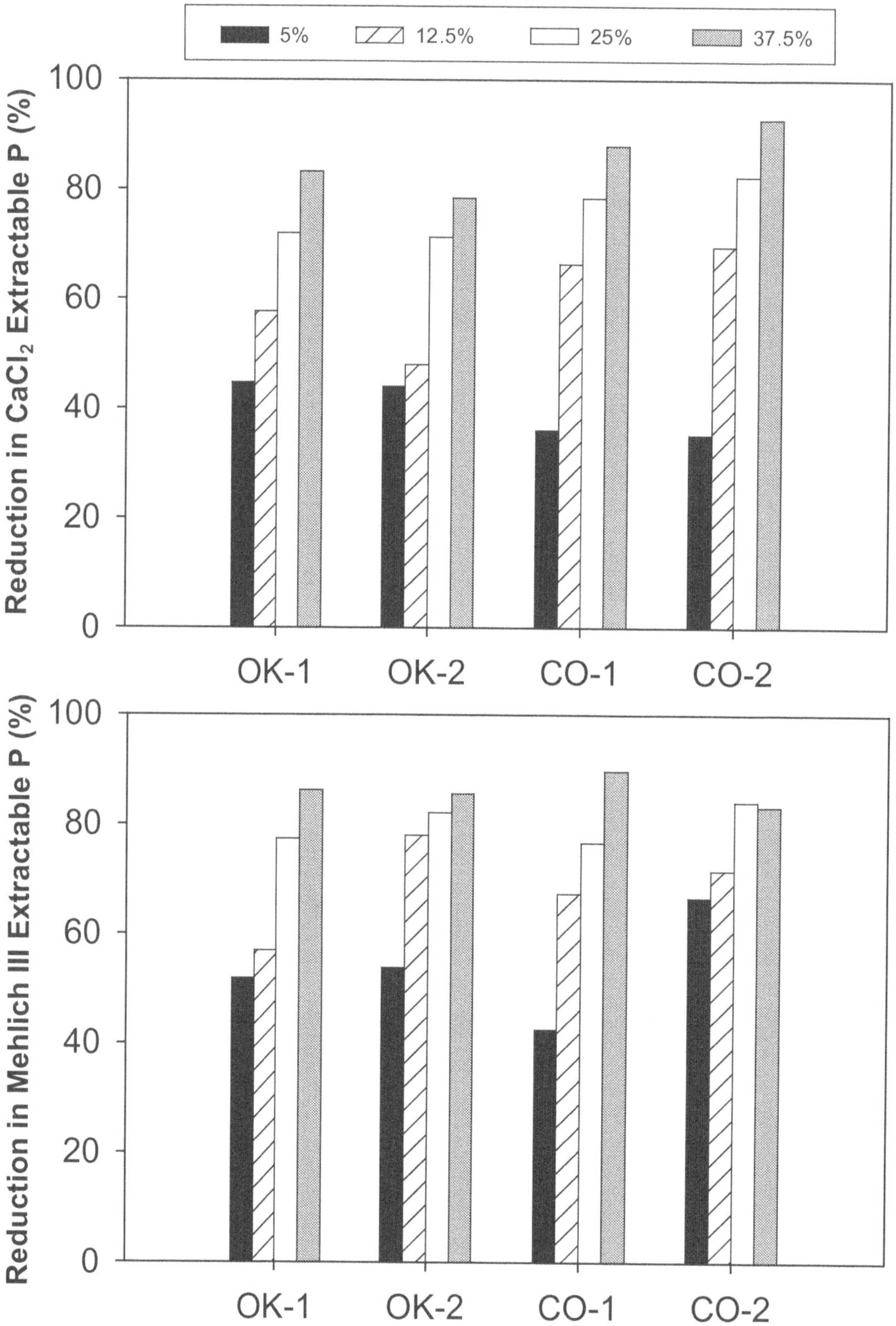

Figure 3.12 WTR blended with biosolids – reduction in P for CO and OK WTRs

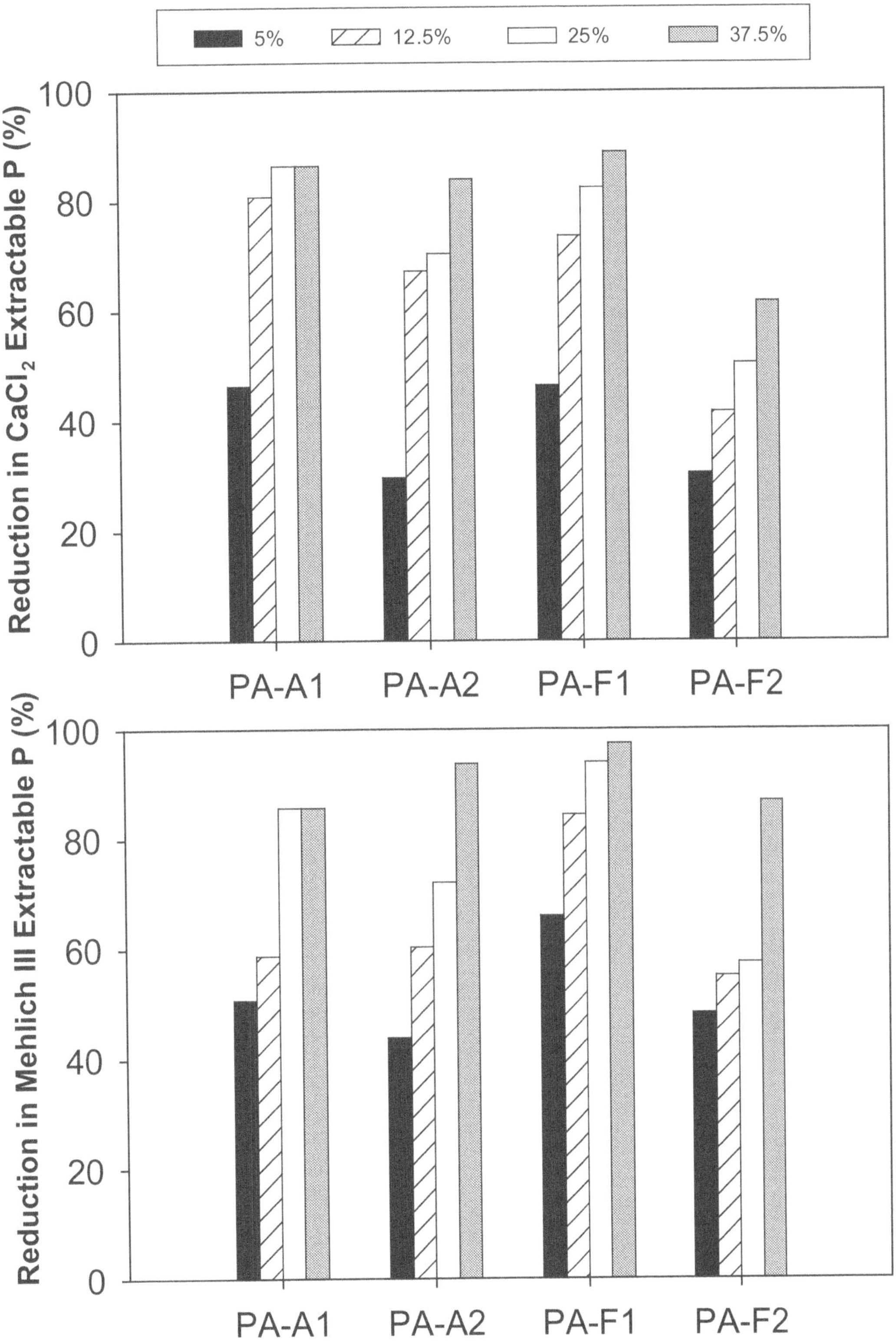

Figure 3.13 WTR blended with biosolids – reduction in P for PA WTRs

PLANT OPERATIONAL DATA CORRELATIONS

Summary of Plant Information

A summary of key operating parameters from the facilities generating the WTRs used in this project are presented in Table 3.9. As a general overview, all WTPs from which these materials were acquired are considered conventional – they utilize coagulation to destabilize raw water particles, clarification to remove the majority of solids formed in treatment, and granular media filtration as a final means of particle removal. However, each of these steps are accomplished slightly differently by each facility's unique infrastructure to accomplish treatment, with varying efficiencies. The level of post treatment used at each facility – chlorination, fluoridation, and pH stabilization (with either caustic or lime) varies between plants, however post treatment was not a major factor in determining the quality of WTR. The major factors that would influence WTR characteristics are:

- The type of coagulant used and the required dosage
- Raw water characteristics that influence the formation and surface chemistry of WTR
  - Inorganic particles – sand, silts, and clays
  - Solid organic materials, such as algae
  - Dissolved inorganics, such as iron and manganese
  - Dissolved organics, such humic and fulvic acids and byproducts of biological activity within the source water
- Other pretreatment additives that contribute to solids generation, such as
  - Polymers (aids for coagulation, flocculation, or filtration)
  - Oxidants, such as potassium permanganate ($KMnO_4$) that react with dissolved materials and form solids
  - Lime used as a pH adjustment (due to the presence of impurities)
  - Powdered activated carbon (PAC, utilized for taste and odor control)

WTPs that directly utilize a river as their raw water source generally have higher raw water turbidities when compared to those drawing raw water from reservoirs. For instance, plants PA-A1, PA-F1 and PA-F2 all had average raw water turbidities of 8.5 NTU or greater. However, it should be noted that the value recorded for OK-1 exhibited an average raw water turbidity of 11.3 NTU. This higher value for OK-1 is likely due to high levels of biological activity within the reservoir. It is known that the quality of this particular reservoir has been compromised due to excessive nutrient runoff within the watershed (as noted in Chapter 1). The impact of a compromised water source on required treatment is also evident in the data. Reservoirs with low turbidity, low total organic carbon (TOC), and low iron and manganese – such as CO-1 and CO-2 – use less treatment chemicals to accomplish finished water goals. Both of these plants use only alum and a polymer coagulant aid on a regular basis. Conversely, the OK-1 plant utilizes substantial quantities of coagulant (38 mg/L alum), coagulant aid (2.6 mg/L), PAC (5.2 mg/L), and is one of two facilities that employs chlorine dioxide (ClO_2) as an oxidant. The PA-A1, PA-F1, and PA-F2 facilities all use river sources, and utilize a broad range of chemicals to produce finished water.

Table 3.9
Summary of WTP operational and physical characteristics for each WTR

WTR Designation	**CO-1**	**CO-2**	**OK-1**	**OK-2**	**PA-A1**	**PA-A2**	**PA-F1**	**PA-F2**
Raw Water Source	Reservoir	Reservoir	Reservoir	Reservoir	River	Reservoir	River	River
Raw Water *								
TOC	2.2	2.2	3.6	2	5	2.4	1.8	2.4
Alkalinity	24	61	115	81	68	16	40	113
Hardness	26	103	140	90	114	33	113	165
pH	7.5	7.9	8.1	7.9	7.6	6.8	8.0	8.1
Turbidity	3.45	2.45	11.3	4	15	1.5	21	8.5
Iron	0.16	0.13	0.19	0.15	0.46	0.14	0.90	0.12
Manganese	0.01	0.02	0.10	0.06	0.07	0.15	0.13	0.17
Finished Water *								
TOC	1.5	1.9	2.3	1.13	2.3	1.4	1.3	0.6
Alkalinity	23	58	123	87	55	20	38	100
Hardness	34	103	137	93	123	46	132	177
pH	7.8	7.8	8.0	8.0	7.1	7.3	6.6	7.3
Turbidity	0.05	0.04	0.09	0.06	0.05	0.03	0.05	0.07
Iron	0.02	0.03	0.03	0.01	0.01	0.07	0.03	0.02
Manganese	0.00	0.01	0.02	0.01	0.01	0.02	0.02	0.02
WTR Form	Solid	Solid	Solid	Solid	Solid	Liquid	Solid	Liquid
% solids	15	20	15	22	18	4	26	4
Dewatering method	Lagoon	SDB	BFP	BFP	BFP	Lagoon	BFP	Lagoon
Age, months	12	6	1	1	1	12	6	12
Treatment Chemicals								
Coagulant	Alum	Alum	Alum	Alum	Alum	Alum	$FeCl_3$	$FeCl_3$
Coagulant dose	22	17.6	38	14	70	15	16	32
Coagulant aid - polymer	1.8	1.6	2.6	1	1.3		1.3	0.4
PAC			5.2	5	5	6	10	
$KMnO_4$					0.5	1.1	0.4	0.4
Chlorine dioxide			1.2					0.4
Pre lime					7	2	13	15

* Values are average for the year samples were collected (2003) in mg/L + SDB – Sand drying bed, BFP – Belt filter press

NOTE: Shaded area in treatment chemicals indicates chemical was not used at the time samples were collected

WTR Constituents

Pretreatment chemical and water quality data were evaluated to determine the relative contribution of each major constituent that comprises WTRs. These major constituents include:

- Solids generated by the coagulant
- Solids removed from the source water
- PAC solids
- Coagulant aid polymers
- Iron and manganese removed
- Solids from impurities in lime (if used in pretreatment)

Solids generated by the coagulant were determined using industry standards (Cornwell et al. 1987) based on the coagulant dose and conversion factors specific to the coagulant used. Solids removed from the source water were estimated using the raw water turbidity, and utilizing a 1:1 conversion factor for determining suspended solids (Cornwell et al. 1987), since source water total suspended solids data were not available for most plants. The amount of solids from PAC and coagulant aid polymers were based on the dose of each chemical, while determining the solids contributed from iron and manganese were based on the difference between raw water and finished water values. It was assumed that lime used in pretreatment contained 10% impurities that contributed directly to solids generation, based on an assessment of MSDS sheets reviewed at the facilities. Solids contributions from polymers used for dewatering were not considered, since normal dosages for dewatering (about 10 lb per ton of dry WTR solids) would account for less than 0.5% of the solids total. Likewise, solids contributions from flocculant and filter aid polymers were not considered, due to the very low dosages of these chemicals (0.07 mg/L or less)

Figure 3.14 shows the relative percent contribution of each major constituent in each WTR. The most readily apparent observation is that in most cases, the make-up of each WTR is unique to a specific location. The exceptions to this are CO-1 and CO-2, which are very similar in the relative distribution of constituents. For each of these WTRs, approximately 65% of the solids contribution comes from the coagulant used during treatment, while approximately 20% of the solids contribution is due to source water turbidity. It is noteworthy that the solids contribution from coagulant aid for both CO-1 and CO-2 - about 15% - was the highest of any of the WTRs evaluated. PAC and lime are not used in the pretreatment process at either of the CO-1 or CO-2 plants, therefore there is no solids contribution from these constituents for these WTRs.

Overall, solids contributed by iron and manganese removal and lime impurities did not substantially contribute to the solids generation; however, the contribution of PAC varied substantially, from about 10% for PA-A1 to over 40% for PA-A2 Overall, the solids contribution estimated from source water turbidity was as little as 10% for PA-A2, while it contributed about 45% of the solids for PA-F1.

The above observations, coupled with the characterization data described earlier in this chapter, indicate that individual WTR characteristics are largely dependent on the unique treatment requirements imposed by source water quality and the performance of the infrastructure utilized for treatment.

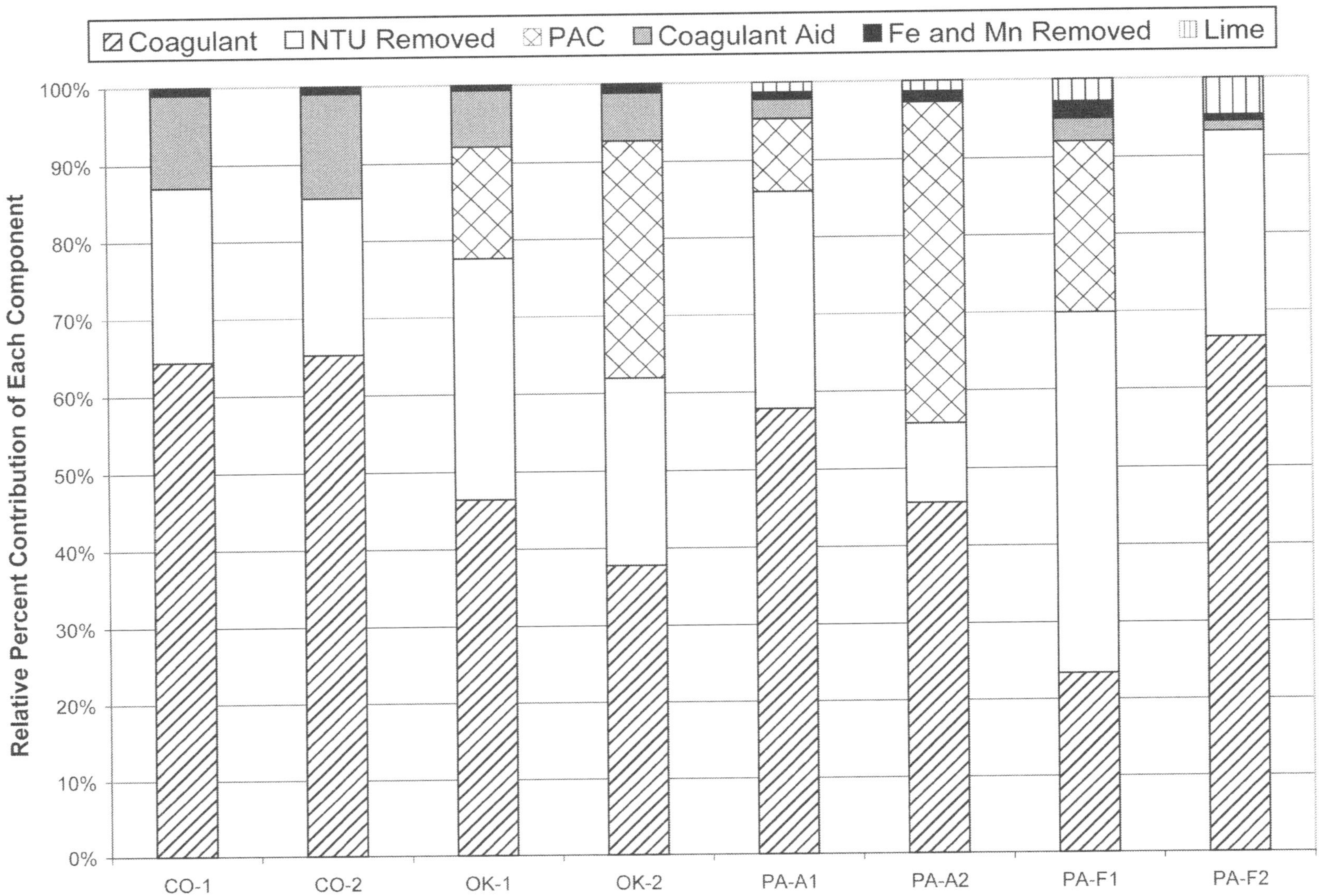

Figure 3.14 Distribution of WTR constituents in each WTR

Correlations with Plant Data

One important goal of the research was to determine if any common operational parameters correlated to the ability of the WTRs to impact P removal. For instance – do high coagulant doses result in WTRs that reduce P more effectively? To evaluate potential correlations, key operational data were compared to research results presented earlier in this chapter.

Since there were strong correlations between Pmax and the impacts on P as determined in the incubation and soil runoff tests, this parameter was compared to the two most likely parameters to have a major impact on the ability of WTR to bind P – coagulant dose and raw water turbidity. The results of these correlations for all WTRs evaluated in this project are shown in Figure 3.15. The data suggest that while general trends may be present, no significant correlations exist between either parameter and Pmax.

Another correlation was evaluated between the ratios of coagulant dose to source water turbidity. This correlation was attempted since many water utilities mainly base their coagulant dose on the destabilization of particles in the source water, which is generally attributed to source water turbidity. The result of this correlation is shown in Figure 3.16. Once again, while a general trend is evident (similar to those in Figure 3.15), no significant correlations exist.

These results strongly suggest that the abilities of WTRs evaluated to bind soil P are not directly attributable to common operational parameters. Rather, the ability to bind P is likely due to a multitude of factors, including but not limited to WTR age, coagulation conditions (pH, temperature, mixing energy, etc.), conditioning during dewatering, and storage time after dewatering or production. Table 3.10 provides a summary of factors and comments related to each.

It should be noted that all attempted correlations were for two parameters. Multivariate and multiple regression analyses were not conducted, because the goal of this research was to establish simple relationships. Future studies should consider multivariate analyses.

Table 3.10
WTR Factors in binding P

Factor	Comments
Dewatering	The greater the amount of dewatering, the WTRs may be less reactive due to reduction in amorphous nature.
Age (after dewatering)	Following dewatering, older WTRs are likely less reactive due to mineralization of aluminum and iron hydroxides complexes.
Coagulations conditions	pH, mixing energies, time of mixing, temperature, type of coagulant, and use of coagulant aids can impact the type of metal hydroxide compounds initially formed in the coagulation process.

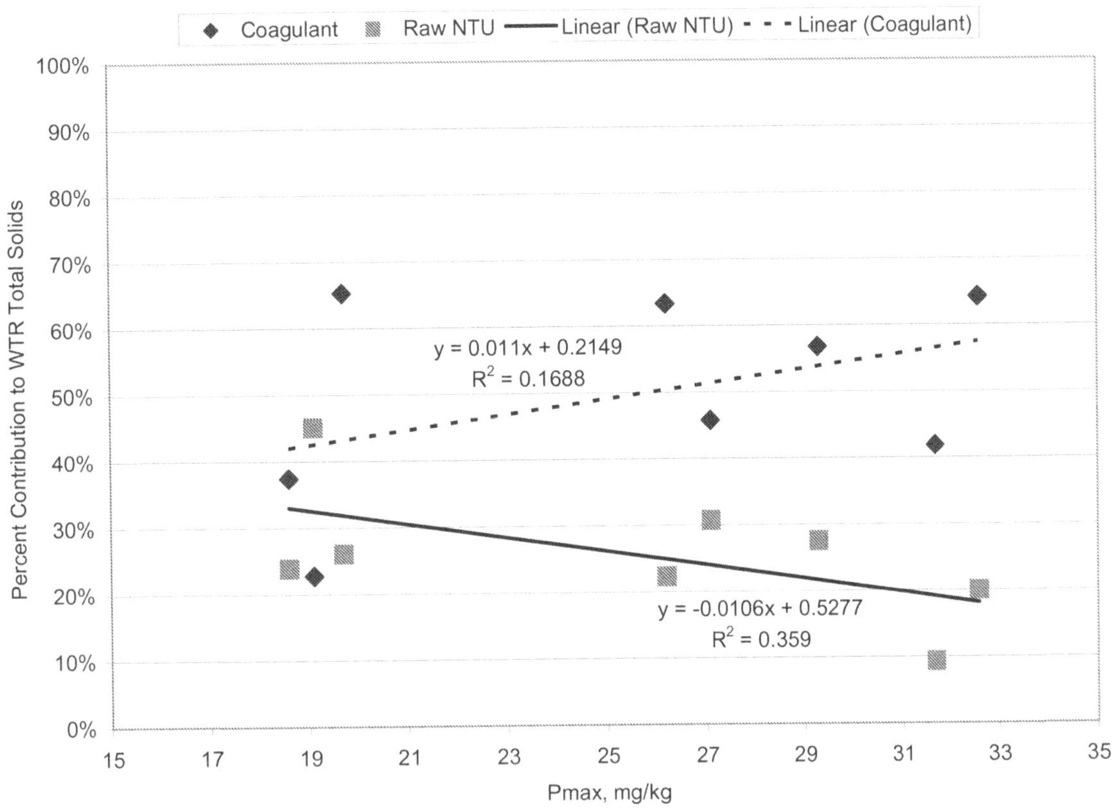

Figure 3.15 Contributions to WTR from coagulant and raw water turbidity versus Pmax

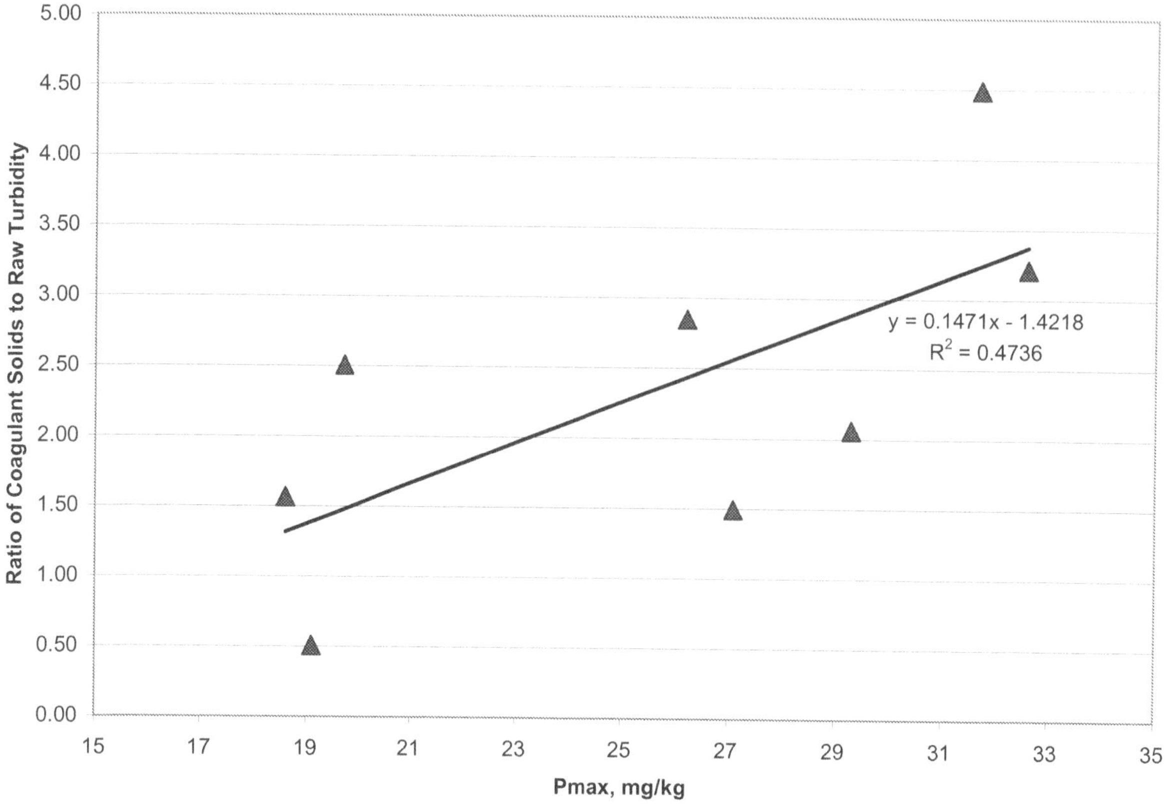

Figure 3.16 Coagulant to raw water turbidity ratio for each WTR versus Pmax

CHAPTER 4
USE OF WTRs IN SOIL NUTRIENT MANAGEMENT

BACKGROUND

As discussed in detail in Chapter 1, substantial focus is now being placed on agricultural P management by a variety of state and federal regulatory agencies including the P level of soils and use of P-bearing soil amendments. A number of efforts are underway to develop best management practices (BMPs) that will assist farmers and land owners in the management of soil P and controlling its impact on water quality.

In response to mounting water quality concerns, many states have developed guidelines for land applying P and watershed management based on the potential for P loss in agricultural runoff (USDA and USEPA 1999). The USDA and EPA created a joint strategy to implement comprehensive nutrient management plans (CNMPs) on animal feeding operations (AFOs) on a national basis by 2008. CNMPs address manure handling and storage, application of manure to the land, record keeping, feed management, integration of nutrient management practices with other conservation measures, and other options for manure utilization.

NRCS state conservationists must decide which of three P-based approaches will be used in nutrient management planning policy. These approaches are:

- Agronomic soil test P recommendations
- Environmental soil test P thresholds, or
- A P Index to rank fields according to their vulnerability to potential P loss.

Many states have already selected the P-based strategies that are used in their Code 590 standard, with the P Index as the basis of most states' strategies.

A major premise of this research was that the use of WTRs could be considered for use in nutrient management programs as a BMP. Most state NRCS agencies have formally adopted the P Index approach as part of their 590 Conservation Practice Standard. This chapter will demonstrate how WTR from each co-sponsor can be applied to their state's current nutrient management program administered through NRCS, and other state agencies that have a role in water quality protection. To the maximum extent possible, the results from the research data presented earlier in the report will be utilized.

It is important to note that the P Index is used by planners and land users to assess the potential risks that exist for P leaving the land and traveling to a surface water body. It also can be used to identify the critical parameters of soil, topography, and management that most influence the movement. Using these parameters, the index then can help select management alternatives that would significantly address the potential impact and reduce the risk. The management alternatives can include BMPs.

To provide a background on how WTR use can be applied as a BMP, a summary of the P Index approaches for each co-sponsor's state is presented in Table 4.1. The table lists the unique source factors, transport factors, watershed conditions and how the P Index is determined in each state.

Table 4.1
Summary of P Index approaches for Pennsylvania, Oklahoma and Colorado

P Index Component	Pennsylvania	Oklahoma	Colorado
P Index trigger	Soil test P > 200 ppm OR Field < 150' to water body	P Index is not currently used; however, nutrient budget required if soil test values exceed crop requirements OR if organic materials are applied	AB-DTPA > 10 ppm Olsen P > 20 ppm Bray-1 P > 30 ppm Mehlich 3 P > 40 ppm AND Water can reach water body
Source Factor			
Soil test P method	Mehlich 3 P	Mehlich 3 P	Olsen P (high pH) Bray-1 P (low pH) Mehlich 3 P
Application rate	lbs P_2O_5/acre per year	lbs P_2O_5/acre per year	lbs P_2O_5/acre per year
Application method	Injection Incorporation Surface applied	Injection Incorporation Sprinkler Irrigation Surface applied	Injection Incorporation Surface applied
Application timing	Season applied Time to application	Time to application	Season applied Time to application
Management	Organic P source availability	No application under certain conditions	BMPs credits allowed that reduce P loss
Transport Factors			
Erosion	Water (RUSLE) *	Water (RUSLE) *	-
Surface runoff class	Soil permeability Field slope	Field slope Depth of soil Rock size and cover	Soil permeability Field slope
Subsurface drainage / flooding	Tile drainage	Flooding Frequency	-
Watershed			
Contributing distance	Distance from field to surface water	Distance from field to water	-
Connectivity	Riparian buffer Direct discharge	Established buffer strip	Field strips Contour buffer strips
Receiving water priority	-	Nutrient limited watershed	-
Index value determination	Multiplicativc	-	Additive

* RUSLE – revised universal soil loss equation

Source: Adapted from Sharpley et al. 2003

PENNSYLVANIA P INDEX

The P Index for Pennsylvania (Weld et al 2003) employs a two-part approach, which is shown in Figure 4.1. The key to this approach is whether traditional nitrogen-based management plans will be applied, or whether P-based plans must be followed. The following description is only a summary of the Pennsylvania P Index; the full document should be reviewed and interpreted by a qualified and certified nutrient management specialist. The approach is described as follows:

- **Part A -** If a field has *either* a soil test P level greater than 200 ppm P (Mehlich-3) *or* is closer than 150 ft. to a receiving body of water, as measured from the lower edge of the field, then continue to evaluate the field using Part B. If neither of these criteria applies, the N-based plan is acceptable as written.
- **Part B** – If either of the triggers for P-based management is met, then the P Index must be used. The P Index for Pennsylvania considers two major factors – P sources and transport.

The following items are considered for developing the source factor:

- **The soil test P level.** A factor of 0.2 is used to multiply the soil test P value as determined by the Mehlich 3 method.
- **Fertilizer P rates.** The method of application and the time of year the fertilizer is applied are considered, and a factor that varies between 0.2 and 1.0 (in increments of 0.2) is used to multiply the fertilizer rate.
- **Manure P rates.** The method of application and the time of year the manure is applied are considered, and a factor that varies between 0.2 and 1.0 (in increments of 0.2) is used to multiply the manure rate. The P availability coefficient of the manure is also considered.

The following items are considered for developing the transport factor:

- **Erosion.** This value is based on the revised universal soil loss equation (RUSLE), and is obtained from the farm conservation plan.
- **Erosion potential.** A value ranging from 0 to 8 in increments of 2 is assigned, and is based on the soil type. Acquired from NRCS nutrient management coordinators.
- **Subsurface drainage.** A value of 0, 1, or 2 is used depending on the existence of an engineered drainage system or the presence of a rapidly permeable soil near water.
- **Contributing distance.** A value ranging from 0 to 8 in increments of 2 is assigned, and is based on the distance to a water body from the lower edge of a field.
- **Modified connectivity.** A factor of 0.7, 1, or 1.1 assigned based on flow path to water

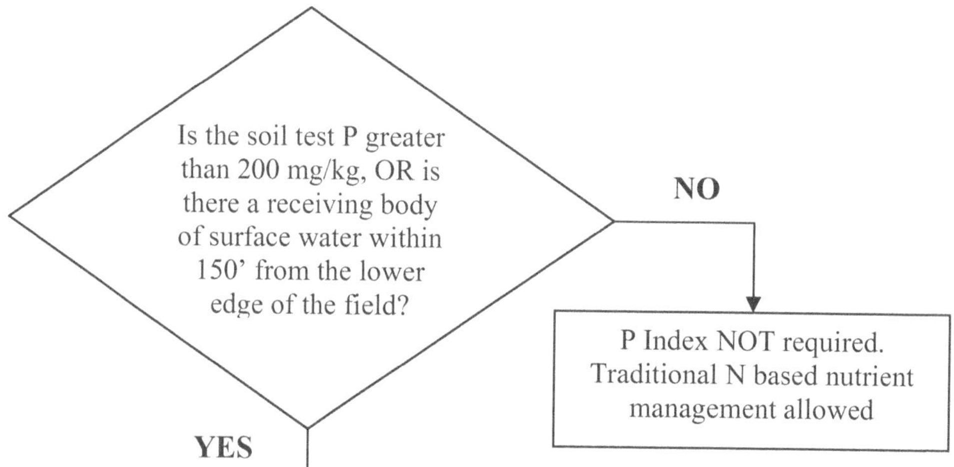

APPLY P INDEX

SOURCE FACTORS

Soil Test Rating = Soil Test x 0.2
Fertilizer Rating = Fertilizer rate x application method factor (0.2 to 1.0)
Manure Rating = Manure rate x application method factor x P availability factor
Source Factor = Soil Test Rating + Fertilizer Rating + Manure Rating

TRANSPORT FACTORS

Erosion = in ton/acre/year based on revised universal soil loss equation (RUSLE)
Runoff Potential = 0 (very low) to 8 (very high)
Subsurface Drainage = 0 (none), 1 (random) or 2 (patterned, or rapidly permeable soil)
Contributing Distance = 0 (>500') to 8 (<150')

Transport Sum = Erosion + Runoff Potential + Subsurface Drainage + Contributing Distance

Modified Connectivity = 0.7 for < 150' w/ riparian buffer, 1 if grassed, 1.1 if direct via pipe

Transport Factor = (Transport Sum x Modified Connectivity) / 22
P Index Value = 2 x Source Factor x Transport Factor

COMPARE P INDEX AND DETERMINE MANAGEMENT GUIDANCE

P Index Value	Rating	Guidance	Action
0 to 59	Low	Apply nutrients to meet nitrogen requirements	Maintain current practices
60 to 79	Medium	Apply nutrients to meet nitrogen requirements	Assess mgmt. & conservation
80 to 99	High	Apply nutrients to meet P crop removal	P based mgmt. plans required
≥100	Very	No P can be applied	P based mgmt. plans required

Figure 4.1 Pennsylvania P Index determination

Once the P Index is calculated, its value is then evaluated to determine the relative rating for the particular agronomic field. Management guidance can vary from maintaining normal nutrient management methods (low rating) to not being able to apply any P to the soil (very high rating). If the guidance requires P based management plans, the land owner must consider measures to reduce the P Index value.

WTR have the potential of reducing the Pennsylvania P Index for P impacted fields in several ways. Table 4.2 identifies how certain factors can be adjusted. Adding WTR to the soil can lower the soil test P value, as demonstrated in this research. If an initial soil test P value is near 200 mg/kg, then the addition of WTR at normal agronomic amendment rates (1 to 10%) could significantly reduce this value, as demonstrated by all PAWC WTRs used in this project. The manure rating also could be reduced, and it is likely that the availability coefficient could be adjusted through the addition of WTR; however, the current Pennsylvania P Index does not recognize WTR as an availability coefficient multiplier. However, because the state is currently considering alum treatment, aluminum-based WTRs may also qualify. Both runoff potential and subsurface drainage factors could be modified with WTR use due to both chemical and *physical* impacts on soils. The addition of clay-like fines present in WTR may help with moisture retention and aggregate stability (Rengasamy, Oades, and Hancock 1980). In permeable sandy soils, the addition of WTR could not only minimize hydraulic conductivity, but also add P sorbing capabilities (Sims, Luka and Eaton 2001; Novak and Watts 2004).

Another potential role for WTR not currently considered in the Pennsylvania P Index could be BMP credits. The addition of BMP credits would largely be dependent on proven performance through full-scale trails performed under rigorous monitoring conditions.

Table 4.2
Pennsylvania P Index factors and the potential impact of WTR

Factor	**WTR Potential in P Management Plan?**	**Comment**
Soil test rating	**Yes**	WTR can be added to soil to reduce the soil test P value
Fertilizer rating	No	Commercial fertilizers are not likely to be modified with WTR to mitigate P availability
Manure rating	**Yes**	WTR can be added to organic fertilizer to either lower the P value, or reduce the availability coefficient.
Erosion	No	The current RUSLE system does not have a variable that could be impacted by WTR
Runoff potential	**Yes**	WTR has the *potential* to modify soil structure through the addition of fine particles which may increase soil aggregate stability.
Subsurface drainage	**Yes**	The addition of WTR to a permeable soil can add both P sorption capability and reduce the rate of hydraulic conductivity through the addition of fine particles.
Contributing distance	No	This distance is fixed depending on site physical features

OKLAHOMA NUTRIENT GUIDELINES

Oklahoma has not adopted a P Index similar to those used in Pennsylvania, Colorado or other states. The Oklahoma NRCS currently uses a version of the NRCS Code 590 (Nutrient Management Standard) to guide nutrient management planning in that state. The Oklahoma version of this document was developed in conjunction with Oklahoma State University, and requires soils in Nutrient Limited Watersheds to be tested once every three years for, at a minimum, pH, nitrogen, phosphorus, and potassium. The results of the test are compared to the requirements of the planned crop.

While Oklahoma has not formally adopted the P Index as standard practice, several elements of the P Index described in Chapter 1 are used to guide the land application of manures or other organic byproducts in the Oklahoma Code 590 standard. Oklahoma NRCS uses an "Oklahoma Phosphorus Assessment Tool" which is a spreadsheet application for easily determining organic byproduct application limits on individual fields. Among its parameters, the tool considers if the field is located within a Nutrient Limited Watershed as determined by the Oklahoma Water Resources Board in Appendix A of the Oklahoma Water Standards. Specific guidelines are provided for organic material (manure) applications based on:

- Soil test P
- Slope of the field
- Depth of soil
- Whether large rocks (>10 inches in diameter) are present.

Table 4.3 provides a summary of the organic material application guidelines based on soil test P levels, field slope, depth of soil, and if the field is located within a Nutrient Limited Watershed. There are other conditions under which no manure or other organic byproducts can be applied, including such factors as distance to streams, excessive slopes (>15%), shallow soils, etc. Oklahoma's current P management tools currently have only two modes under which WTR can directly be used:

- Soil incorporation to reduce the soil test P value.
- Blending with organic byproducts prior to application.

The Oklahoma Code 590 recognizes the ability of organic materials to improve physical, chemical, and biological condition of soil; however, no overt recognition is currently given to materials that help reduce the pollution potential of P or other constituents. The code does require that plans and specifications for nutrient management identify when a field will become P limited, to include discussion about how P accumulation will be addressed, and what measures will be taken to minimize P pollution. Further, other codes are referenced that, while not specifically included in the nutrient management code, could also incorporate WTR. These other codes include standards for contour buffer strips and conservation waterways. WTR could be incorporated in soils in grassed areas to enhance the ability to reduce soluble P runoff from fields.

Table 4.3
Annual organic material application rates for OK

Rating	Soil Test P, mg/kg (Mehlich 3 Analysis)		Organic Material Application Rate Guidance Based on Slope and Soil Depth *			
	Non Nutrient Limited Watershed	Nutrient Limited Watershed	0 to 8% Slope Soil > 20" Deep	8 to 15% Slope Soil > 20" Deep	0 to 15% Slope Soil 10" to 20" Deep	Rocks >10" dia. that cover >3% of the surface and <8% slope
Low	0 – 65	0 – 65	Full Rate	Full Rate Split Application	Half Rate	Half Rate
Moderate	66 – 250	66 – 120	Full Rate	Half Rate	Half Rate	Half Rate
High	251 – 400	121 – 300	Half Rate	Half Rate	Half Rate	Half Rate
Very High	> 400		Plant Removal	Plant Removal	Plant Removal	Plant Removal
Severe	#	> 300	No Application	No Application	No Application	No Application

Source: USDA 2004

* Organic Material Application Rates

- Full Rate – Not to exceed the N requirement of the crop and the following P_2O_5 rates:
  - 200 lbs P_2O_5 per acre when surface applied.
  - 300 lbs P_2O_5 per acre when application is by sprinkler irrigation and managed to prevent runoff from field.
  - 400 lbs P_2O_5 per acre if injected below the soil surface or surface applied and incorporated within 7 days.
- Half Rate – Not to exceed the N requirement of the crop and the following P_2O_5 rates:
  - 100 lbs P_2O_5 per acre when surface applied.
  - 150 lbs P_2O_5 per acre when application is by sprinkler irrigation and managed to prevent runoff from field.
  - 200 lbs P_2O_5 per acre if injected below the soil surface or surface applied and incorporated within 7 days.
- Split Application – Not to exceed the N requirement of the crop
  - Application no more than ½ the allowed P_2O_5 rate per application at least 30 days apart.

Severe Rating - No Application

COLORADO P INDEX

The Colorado P Index (COPI) Risk Assessment was developed in cooperation with Colorado State University Cooperative Extension (CSUCE) to comply with the NRCS Practice Standard, Code 590 (Sharkoff, Waskom, and Davis 2003). Similar to the versions used in other states, a field-by-field risk assessment of the potential for P transport is required for all nutrient management plans that include land application of organic byproducts.

A Preliminary P Risk Screening Tool is included with the COPI, and intended to make an initial determination for fields that meet the P risk assessment requirement. If the preliminary screening tool indicates that manures or other organic by-products may be applied according to crop nitrogen requirements, the field specific information is documented and includes: 1) soil test P, 2) extraction method, 3) sample depth, and 4) reason why storm water runoff and/or irrigation tailwater cannot reach a surface water resource. This information is then included in the plan to meet the risk assessment requirement. Nutrient management plans that include land application of manures or other organic by-products must also include the following:

- The land base requirement needed to enable plan implementation based on P, even when initial implementation will be based on nitrogen. This is a unique feature that requires land managers to consider alternatives before they are required, essentially helping them consider a "what-if" scenario.
- The soil test P level and corresponding extraction method at which plan implementation on a P basis should be implemented. Again, this requirement gets land mangers to consider alternatives.
- Information about conservation practices and management activities that can decrease the potential for P movement from the field. Documentation is also required to show that field-by-field P risk assessment results were discussed with the person managing the field. This feature helps to create an accountability aspect for the Colorado P Index.

Figure 4.2 is the decision tree that is used for the preliminary assessment of fields that may be required to use the Colorado P Index. Similar to Pennsylvania, certain conditions exist that would allow conventional nutrient management plans to be used:

- Organic materials are not applied
- Soil test P levels are below certain values. The values are dependent on the soil P test used, which is dependent on soil pH and other factors.
- Storm water runoff or irrigation tailwater can not reach a nearby water body.

The Colorado Phosphorus Index consists of four site and management Risk Factors that can affect the potential for the movement of P off-site. In order to complete an assessment the relative risk associated with each of these four Factors must be rated. Once they are rated, the values are added together and compared with the Risk Interpretations to determine the relative potential for the movement of P off-site. Implementing certain Best Management Practices may also be considered to mitigate or decrease the relative risk potential. The four Risk factors include:

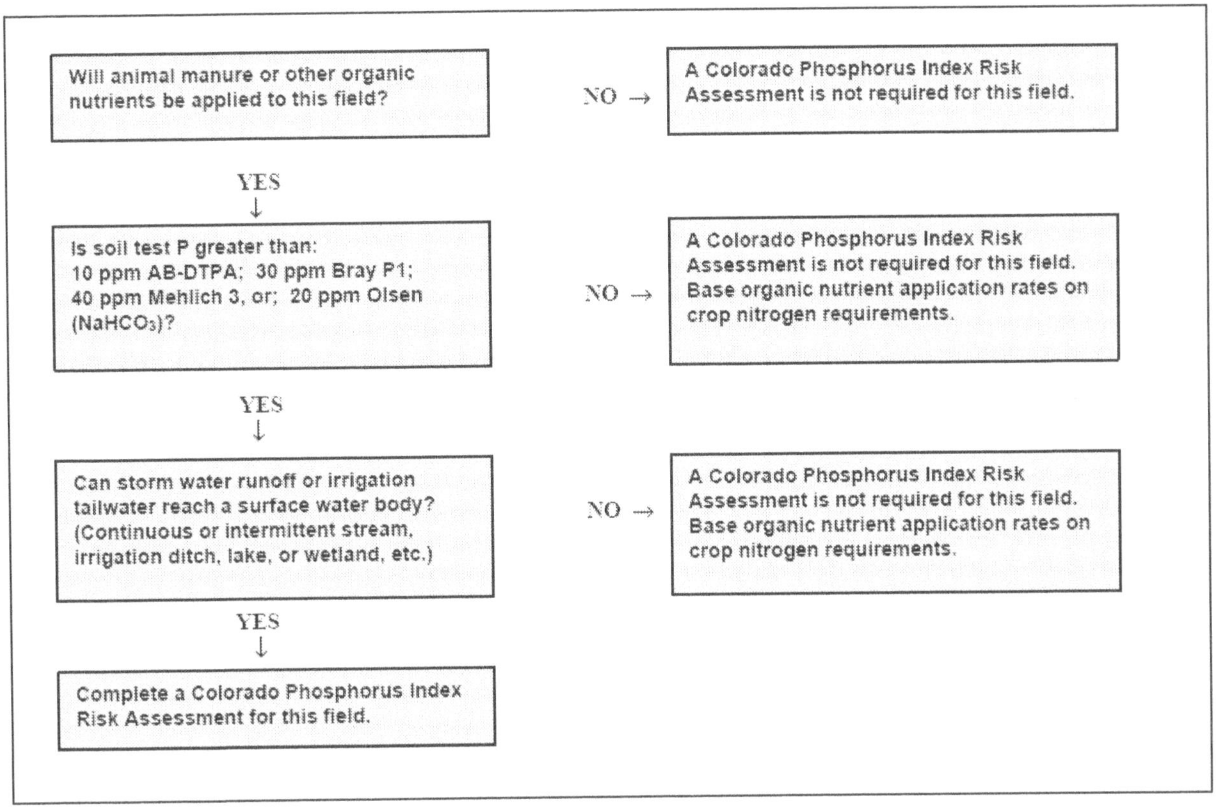

Source: Sharkoff, Waskom, and Davis 2003

Figure 4.2 Colorado preliminary P Risk screening tool

- **Runoff class** – Runoff Class is based on field slope and the least permeable soil layer in the top three feet of the soil profile. Permeability classes for specific soils are based on Soil Survey information. Soil permeability class and field slope must be determined first, and then the runoff class risk can be determined. The value assigned ranges from Low (1) to Very High (4).
- **Soil test P** – Soil samples are acquired using tests based on soil pH and calcium carbonate content. There are specific sampling technique requirements depending on land use. Based on the results, a value is assigned from Low (1) to Very High (4).
- **P application rate** – The P application rate is the amount of P (P_2O_5 basis) annually applied (or average annual application rate calculated for the current crop rotation) to the field in pounds per acre from both inorganic and organic sources. The rate is based on specifications for manufactured fertilizers, and on accepted values for organic materials depending on the source (cows, pigs, sheep, etc.). The amount of P applied is then used to assign a value from Low (1) to Very High (4).
- **P application method** – A value from Low (1) to Very High (4) depending on the method of application (surface applied, incorporated, or via sprinkler) and the time of year.

- **Best Management Practices (BMPs) Implementation Credits** – Colorado has appropriately recognized the value of BMPs and has incorporated their use in its P Index. Specific BMPs are allowed to decrease the relative potential for off-site P movement. When BMP credits are used, it allows for one point to be subtracted from the gross score for *each one* of the following BMPs:

  - Contour Buffer Strips
  - Cover Crops
  - Filter Strips
  - Furrow Diking
  - Grassed Waterways
  - Linear polyacrylamide
  - Residue Management
  - Terraces

The index does *not* prescribe what dose of polyacrylamide (PAM – a compound that can increase the water holding capacity of soils) should be applied. When conducting watershed or sanitary surveys, Colorado utilities should seek to identify if PAM is being used in waters that could eventually drain into raw water sources. PAM could have direct impacts on drinking water treatment coagulation processes due to the contribution of negatively charged particles.

The individual scores for each risk factor are then added to determine the COPI overall score. Table 4.4 has a summary of the management requirements based on the COPI overall score.

Table 4.4
Management requirements associated with Colorado P Index (COPI) overall score

Net Score	Risk Interpretation
<8	This field has a **LOW** potential for off-site P movement if management is maintained at the current level. Organic nutrient application rates may be calculated according to crop nitrogen requirements.
8 to 11	This field has a **MEDIUM** potential for off-site P movement and some management changes may be required to support continued long term organic nutrient applications. Organic nutrient application rates may be calculated according to crop nitrogen requirements.
12 to 15	This field has a **HIGH** potential for off-site P movement and management changes should be implemented to decrease risk. Organic nutrient application rates should be calculated according to crop P requirements.
16	This field has a **VERY HIGH** potential for off-site P movement and management changes are needed to decrease risk. Organic nutrients should not be applied to this field.

Source: Sharkoff, Waskom, and Davis 2003

WTR have the potential of reducing the COPI in several ways. Table 4.5 identifies how certain factors can be impacted. Adding WTR to the soil can lower the soil test P value, as demonstrated in this research. If an initial soil test P value is near 200 mg/kg, then the addition of WTR at normal agronomic amendment rates (1 to 10%) could significantly reduce this value, as demonstrated by all DW WTRs used in this project. The fertilizer rating also could be reduced for organic materials that are pretreated. While runoff class risk is not likely to be directly impacted by WTR, the COPI is set up to favor infiltration. Therefore, if WTR is used routinely to lower soil P, soil permeability testing should be performed routinely to determine if any detrimental impacts on infiltration have occurred due to the use of WTR. Finally, the COPI already has a system in place to recognize BMPs. Therefore, there is strong potential that land applied WTR could be incorporated to provide BMP credit after rigorous field testing has occurred to prove its benefits in runoff P control.

Table 4.5
COPI factors and the potential impact of WTR

Factor	**WTR Potential in P Management Plan?**	**Comment**
Runoff class risk	No	Colorado favorably views high permeability of soils. Therefore, if WTR are used to reduce values of other factors, soil permeability should routinely be tested to ensure no detrimental impacts have occurred due to WTR use.
Soil test P risk	**Yes**	WTR can be added to soil to reduce the soil test P value
P application risk	**Yes**	WTR can be added to organic fertilizer to either lower the P value, or reduce the availability coefficient.
P application method	No	WTR would have no effect on the method of application or time of year that fertilizers are applied
BMPs	**Yes**	Although not currently listed, WTR use could be added to the approved list of BMPs after rigorous field testing has occurred to prove its benefits in runoff P control.

SUMMARY

Each co-sponsor's state has unique requirements for managing soil P with the intent of protecting surface water quality. Each state's requirements are based on numerous factors specific to regional features, including but not limited to soil characteristics, meteorological conditions, farming practices, and topography. Some P indices already have pathways for the use of WTR in helping manage soil P, while others may require development.

CHAPTER 5
CONCLUSIONS AND RECOMMENDATIONS

CONCLUSIONS

Project conclusions are discussed below. They are categorized based on the major areas of focus performed throughout the project.

Extended Characterization Value

The results of extended characterization demonstrated that each WTR has unique chemical characteristics that impact the ability of each WTR to react with P. These unique characteristics are attributable to the following:

- Type of coagulant used
- Dose of the coagulant
- Source water constituents from in-situ sources
- Direct impacts of chemical properties of treatment chemicals
- Indirect chemical properties of treatment chemicals
- Storage conditions
- Age of water treatment residuals
- Geology of physiographic region that impacts properties of soil particles present in source waters

There are several tests that did provide valuable insight into the ability of WTR to bind P, which correlated well with research results, the most significant being Pmax (the maximum amount of P that is capable of being adsorbed to WTRs). There was a high degree of correlation between Pmax and the concentration of reactive aluminum and iron as measured by ammonium oxalate extraction. The testing also revealed that several WTRs possessed substantial levels of major soil nutrients, including nitrogen, calcium and potassium. The presence of these nutrients is likely due to source water conditions; however, the presence of nitrogen could also be attributed to the use of amine based polymers as coagulant, flocculant, filter, or dewatering aids. Plant micronutients – manganese, selenium and others - may also be present.

P Reduction When WTR Incorporated in High P Soils

Tests evaluating the blending of WTR with a high P soil demonstrated that loading rates between 1 and 10% (0.2 and 2.2 Mg/ha) were able to significantly reduce both soil test P and soluble P. The following conclusions were possible based on the test results:

- All WTRs had varying degrees of impact on both soil test P and soluble P. This helps to defend the general thesis that each WTR possesses unique P adsorption characteristics.
- All WTRs were able to reduce soil test P to less than 200 mg/kg at the 10% loading rate, and in most cases were able to do this after just one week of incubation time.

- The majority of the soil test P reduction occurred after only one week of incubation. This is likely due to the large surface area available on the solids surfaces of both the WTRs and the high P soil.
- Of the WTRs tested, the most reactive was PA-F1, an iron coagulant-based WTR; however, general assumptions should not be made about the relative sorption capacity of iron based-coagulants compared to aluminum based-coagulants, since only one iron coagulant based WTR is included in the full range comparison.

WTR that is stored in lagoons and directly land applied without dewatering prior to application exhibited substantially greater ability to bind P when compared to the same WTR that had been dewatered. This was observed for both liquid WTRs evaluated in this research. This difference is likely due to the more amorphous and reactive nature of the aluminum and iron hydroxides (less crystalline and mineralized) when they are in the liquid form, along with the likelihood that the wet materials have a greater surface area than when they are dried. The maximum amount of liquid WTR that can be applied to a soil will likely be related to the hydraulic capacity of the soil such that runoff does not occur from the application. Significant correlations (R^2 values of 0.61 to 0.86) were observed between Pmax and the percent reduction in both soil test P and soluble P. This helps to confirm the value of this parameter for guiding the selection of WTRs for beneficial use as a P binding soil amendment.

The ability for WTR to bind P must be taken into account when evaluating a drinking water utility's overall residual management philosophy. The location of the treatment plant, the nature and P sorption capabilities of its WTRs, distance to agronomic areas needing P control, cost of residuals storage, transportation costs, and other disposal options all must be factored to arrive at the most optimal solution for both the utility and the environment.

P Reduction in Runoff When WTR Surface Applied

Simulated rainfall testing also confirmed the ability of WTRs to reduce P in runoff for soils that are high in P and have also recently received organic materials (poultry manure). The addition of liquid WTRs caused a further decrease in runoff P. The WTR applied as liquid reduced P slightly better than when dried. This is likely attributable to the fact that liquid WTRs have a greater reactive oxide surface area than when dried, which is consistent with observations made in the high P soil blending tests. As with the incubations results, strong correlations (R^2 values of 0.52 to 0.92) exist between the Pmax values determined during the initial characterization and the results of the simulated runoff tests for the six solid WTRs.

Other factors including storage costs, dewatering costs, distance to use, differential performance between wet and dry WTR, and transportation costs must all be considered when evaluating residuals management options.

The majority of P runoff from the test plots occurred after the initial rainfall event. This further confirms the potential benefit of co-applying WTR at the same time organic material is applied (to gain the greatest benefit of its P sorption potential). The research evaluated WTR loaded in a concentrated strip (25% of surface area), which prior research (Basta and Storm 1997; Sims, Luka, and Eaton 2001; Miller et al. 1997; Jacobs and Teppen 2001; Daniel et al. 1999) has shown to be highly effective for intercepting soluble P. It is reasonable to assume there is a strong potential that WTR loaded on greater areas of agronomic fields, or incorporated with the organic material prior to application, could further reduce soluble P.

P Reduction When WTR Blended with Organic Materials

Tests that blended WTR with biosolids and poultry manure revealed that 1) WTR were able to substantially reduce both soil test P and soluble P, and 2) the percent reduction was directly related to the amount of WTR used. Generally, soluble P levels were substantially reduced by all WTRs at loading rates at or above 25% ratio (of WTR to organic material). Also, most WTRs exhibited little or no incremental increased impact on soluble P above a 25% loading. It should be noted that substantially high blending ratios (WTR : organic material) would require large amounts of WTR, which may not be practical from an availability perspective; however, the intent of WTR addition is not to reduce the fertilizer value of organic materials, but rather to control the rate at which P is released – especially after initial application. This aspect was not evaluated in this research.

WTR Condition Considerations

The experiments conducted for this research utilized WTR that had been processed to a uniform size. This was done to help ensure more precise correlations and observations between the various WTRs that were evaluated. Further, the moisture content and temperature of the mixtures used for the WTR/soil and WTR/organic material incubations were kept nearly constant. Substantial differences may occur when unadulterated WTR acquired directly from a water treatment facility is land applied.

RECOMMENDATIONS

The following recommendations are offered to the co-sponsoring utilities that supported this tailored collaboration project. These recommendations also should be considered by other water utilities that are interested in pursuing the beneficial use of their WTRs for the control of P in soil environments and the protection of surface water quality. The sequence of steps for considering WTR as a BMP for NPS P pollution control is shown in Figure 5.1.

WTR Characterization

Utilities should maintain ongoing and current characterization of their respective WTRs. This characterization should include all the parameters specified by the state regulatory agencies that oversee the management of WTR, together with parameters that determine the ability of the WTR to bind P. These parameters include Pmax, as determined through isotherm testing, and reactive aluminum and iron, as determined through ammonium oxalate extraction or similar methods.

Currently, only research-level laboratories at universities and private industry posses the ability to conduct isotherm testing to develop Pmax values that this research showed to be highly predictive of their ability to reduce P pollution potential. Constructing P sorption isotherms are time consuming and labor intensive, and will thus have little widespread use by analytical laboratories. Another option might be to conduct a single-point P sorption Index. This method is used in soil testing to determine P retention on high P fixing volcanic soils (Bache and Williams 1971). Alternatively, Mehlich 3 extractable soil aluminum and iron could be used as an estimate of P sorption. This method could be useful if Mehlich-3 P is already being determined and the relevant analyses can be conducted by ICP on the one extractant (Kleinman and Sharpley 2002).

However, the science of soil testing is advancing, and many soils laboratories are now developing the capability to provide advanced testing services. Therefore, each utility is encouraged to identify and establish routine dialogues with laboratories that are developing or have established these advanced testing methods.

Advanced analyses can be coupled with routine testing that is required by state regulatory agencies. The differences in WTRs observed during this research support the recommendation that utilities characterize their WTRs on a seasonal basis due to changes in source water quality or the required level of pretreatment chemicals. For instance, spring conditions (whether due to snow melt or spring rains) can create higher raw water turbidities and may have lower alkalinity and hardness, yet require greater amounts of coagulants and other chemicals used for particulate removal. Also, powdered activated carbon (PAC) use may be greater in the summer or early fall to address taste and odor issues due to algal activity in the source water.

California-based utilities should confirm these results using the California Waste Extraction Test, a more aggressive leaching procedure that is used in that state to characterize hazardous waste.

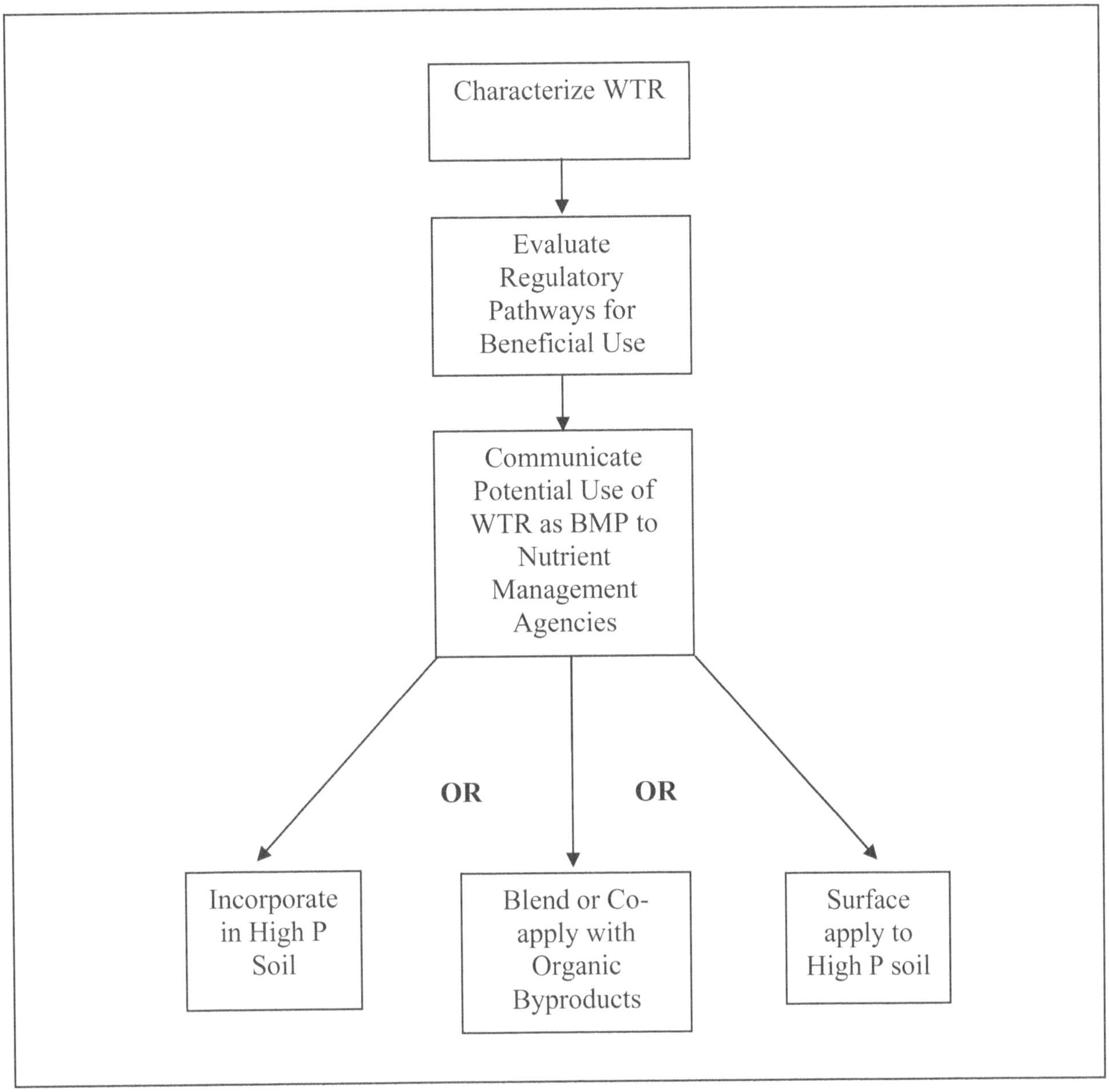

Figure 5.1 Process to consider WTR for beneficial use to control NPS P pollution

Communicate Results to Nutrient Management and Regulatory Agencies

Each of the co-sponsors, and other drinking water utilities that are interested in using WTR as a BMP for soil P pollution control, should actively pursue relationships with local NRCS and related agencies to communicate the potential role of WTR in managing P in agronomic soils and organic materials management programs, especially those involved with the management of CAFOs. Utilities should also communicate the results of this research to state agencies that are responsible for regulating the management and fate of WTR. This is especially true for states that have emerging WTR regulations, WTR regulations that are being reviewed, or established beneficial use programs.

Pennsylvania

Pennsylvania regulates WTR through its Bureau of Waste Management, Division of Municipal and Residual Waste under Article IX, Chapter 287. While this provides general requirements for WTR, land application of WTR is further defined in Ch. 291; however, a Beneficial Use program is also administered, and several general permits already exist for the land application of WTR. These permits should be reviewed to determine if P control can be adopted within the conditions, and to determine if a completely new general permit specific to soil P management is warranted. Further, Pennsylvania's biosolids regulations, administered through the Bureau of Water and Wastewater Management (found in Ch. 271 of the Pennsylvania Code) should be reviewed and appropriate dialogues established to determine the role WTR can play in the land application of biosolids.

The PA DEP Bureau of Watershed Management also should be contacted regarding the potential role WTR can have in their programs. The Pennsylvania Association of Conservation Districts (PACD) provides mini-grants for the demonstration of technologies that can be used for NPS pollution control. This funding is provided through the Chesapeake Bay Foundation and the EPA's Section 319 program (PACD 2003). The Bureau sponsors a NPS Management Program to coordinate efforts to address NPS pollution and watershed restoration.

Colorado

Colorado has recently revised WTR regulations for WTR management, administered through the CDPHE Water Quality Control Division (CDPHE 2004). This regulation does contain provisions for beneficial use, which require the submittal of a Beneficial Use Plan. The co-application of WTR with biosolids is regulated under Colorado's Biosolids Regulations; therefore the conditions of these regulations also should be reviewed. It should also be noted that the presence of radionuclides is an emerging issue for WTR management in Colorado, prompted by the Radionuclides Rule (EPA 2000) and issuance of draft guidance to address WTRs that may contain radionuclides.

Formal dialogues to address these issues have been initiated through the Rocky Mountain (RM) Section AWWA Water Utility Council (WUC). The RM Section WUC is encouraged to utilize the results of this research to show that substantial environmental benefits can be gained through the direct land application of WTR, or co-application with biosolids, and the loading rates utilized will likely not have a substantial impact on background radiation levels of unamended soils.

Oklahoma

Oklahoma does not have formal WTR regulations, but has been at the forefront of activities to address P pollution in watersheds recognizing their substantial impacts on drinking water supplies. In fact, one of the two demonstration watersheds used nationally for implementing federal nutrient management requirements was the Lake Eucha Watershed. The degraded water quality within the watershed and Lake Eucha has been directly linked to the runoff of P-laden manures from densely concentrated poultry-growing operations in the watershed. Oklahoma utilities, and especially TMUA, should communicate the results of this research to lawmakers that are finalizing regulations for the management of organic materials from CAFOs.

Utilize NRCS Programs and Codes

BMPs that utilize WTR for NPS P pollution must control the delivery of P to receiving water resources by:

- Minimizing P availability (source reduction),
- Reducing the transport and/or delivery of P, and
- Intercepting P in runoff before it is delivered to the water resource.

In addition to Code 590 for Nutrient Management, the NRCS has numerous National Conservation Practice Standards that have been developed to guide specific activities related to agronomic practice. Some of these Standards could incorporate WTR to accomplish the goals specific to each standard. A summary of the NRCS Standards that could potentially use WTR are shown in Table 5.1. For instance, the use of WTR in vegetative buffer strips has been shown to control runoff P (Eaton and Sims 2001). The table lists the Code number assigned to each Standard, its title, definition, and a brief description of the role WTR could play in that Standard. Each co-sponsoring utility is encouraged to review these Standards.

Dialogue with state NRCS and related agencies (described above) should include a review of these Standards. It also is important that the use and application of WTR in these Standards does not reduce or eliminate the requirements set forth by state regulatory agencies for the management of WTR; however, there may be a strong potential for utilities to justify proposed beneficial uses when the declared beneficial use can be directly associated with an existing NRCS Standard.

It is very important to note that application of WTRs to sequester P in high P soils or organic byproducts must be associated with a reduction in future application rates of P (in manure or biosolids). Adding WTRs does have strong potential as a BMP, but like all the others is a stop gap measure until input reductions are achieved.

WTR Processing

The results of this research showed that liquid WTRs – those stored in lagoons or similar structures – are more reactive for binding P when compared to the same material after it has been dried. Therefore, utilities that use lagoons should consider this fact when considering the ultimate fate of WTR from storage units when they are cleaned to provide capacity for continued operation. Further, utilities that are considering modifying residuals management programs for greater capacity should include the enhanced ability of liquid WTR to bind P as a factor when selecting preferred management methods. Since watershed activity has a major influence on source water quality and the required treatment to accomplish finished water goals, it is evident that watershed protection should be pursued to the maximum extent possible.

This research was conducted using processed WTR to ensure a uniform size and to better evaluate the differences in adsorption characteristics between WTRs. It is highly likely that there will be a considerable range in actual particle sizes removed from a WTP and land applied. This fact will likely have a substantial impact on the reaction rates with both soil P and soluble P. Therefore, field studies are highly recommended to better discern the benefits of land applied WTR to control P loss and protect surface water quality.

Supplemental Measurements

WTR will likely only decrease P loss in runoff when erosion is controlled. Application of these amendments to bare soil may have little effect on mitigating runoff P associated with erosion. For example, dissolved P may be decreased by P being sequestered by the added WTR but particulate losses due to high rainfall and runoff energy would have little effect on total P losses. If total P losses are not reduced then this could be a major concern to mitigating water quality impacts of high P soils. Likewise, blending WTR with soil and organic byproducts reduces both soil test and soluble P as demonstrated in this research; however, there may be concerns with the fate of P that becomes bound to WTR and enters the aquatic environment in a particulate form. Therefore, both dissolved and total P in runoff should be measured when erosion is suspected.

Table 5.1
Summary of NRCS Conservation Practice Standards and potential role of WTR

Code Number	Title	Definition	Role of WTR
311	Alley Crops	Trees or shrubs planted in a set or series of single or multiple rows with agronomic, horticultural crops or forages produced in the alleys between the rows of woody plants.	WTR can be used to help establish alley crops due to enhanced moisture retention and other potential physical and nutrient benefits
322	Channel Bank Vegetation	Establishing and maintaining vegetative cover on channel banks, berms, spoil, and associated areas.	WTR can be incorporated into vegetative strips or routinely applied to aid in the interception of soluble P in runoff
332	Contour Buffer Strips	Narrow strips of permanent, herbaceous vegetative cover established across the slope and alternated down the slope with parallel, wider cropped strips.	WTR can be incorporated into buffer strips or routinely applied to aid in the interception of soluble P in runoff
360	Closure of Waste Impoundments	The closure of waste impoundments (treatment lagoons and waste storage ponds), that are no longer used for their intended purpose, in an environmentally safe manner.	WTR can be incorporated with fill or impoundment materials to reduce P pollution potential. If impoundment is converted for other use, WTR can help mitigate P impact
386	Field Border	A strip of permanent vegetation established at the edge or around the perimeter of a field.	WTR can be incorporated into border or routinely applied to aid in the interception of soluble P in runoff
393	Filter Strips	A strip or area of herbaceous vegetation situated between cropland, grazing land or disturbed land (including forestland) and environmentally sensitive areas.	WTR can be incorporated into filter strips or routinely applied to aid in the interception of soluble P in runoff
412	Grassed Waterway	A natural or constructed channel that is shaped or graded to required dimensions and established with suitable vegetation.	WTR can be incorporated into soil or routinely applied in grassed waterways to aid in the interception of soluble P in runoff

(continued)

Table 5.1 (Continued)

Code Number	Title	Definition	Role of WTR
422	Hedgerow Planting	Establishment of dense vegetation in a linear design to achieve a natural resource conservation purpose.	WTR can be used to help establish woody plants due to enhanced moisture retention and other potential physical and nutrient benefits
521B	Pond Sealing or Lining	A liner for a pond or waste impoundment consisting of a compacted soil-dispersant mixture.	WTR could be used in place of or in conjunction with liner material to reduce permeability *and* to bind P in impounded water
572	Spoil Spreading	Disposing of surplus excavated materials.	WTR can be incorporated with spoils that are high in P to reduce pollution potential, or they can be used to enhance physical structure
590	Nutrient Management	Managing the amount, source, placement, form, and timing of the application of plant nutrients and soil amendments.	WTR can be directly applied to land surface or incorporated with plow layer of soil to bind P, or can be blended with organic byproducts to reduce applied P
601	Vegetative Barrier	Permanent strips of stiff, dense vegetation along the general contour of slopes or across concentrated flow areas.	WTR can be incorporated into vegetative strips to aid in the interception of soluble P in runoff
612	Tree / Shrub Establishment	Establishing woody plants by planting seedlings or cuttings, direct seeding, or natural regeneration.	WTR can be used to help establish woody plants due to enhanced moisture retention and other potential physical and nutrient benefits
633	Waste Utilization	Using agricultural wastes such as manure and wastewater or other organic residues.	WTR can be added to wastes to reduce the coefficient of P availability
635	Wastewater Treatment Strip	A treatment component of an agricultural waste management system consisting of a strip or area of herbaceous vegetation.	WTR can be added as pretreatment to wastewater, or to vegetative strips to aid in the interception of soluble P in runoff

FUTURE RESEARCH NEEDS

The results produced from this tailored collaboration project identified several items that require further investigation to further define the role WTR can play as a BMP for the control of P loss from agronomic soils to protect water quality.

- Controlled field demonstrations should be conducted to better translate the impact of WTR for soil P control. The field demonstrations could occur on a small or large scale, depending on the available resources. The larger the scale of the demonstration, a greater number of samples will have to be acquired.
- This research was conducted using processed WTR to ensure a uniform size and to better evaluate the differences in adsorption characteristics between WTRs. Together with field studies, the significance of WTR particle size as it relates to the both the rate of reaction with forms of P and the overall capacity (Pmax) should be evaluated.
- Development of a simple predictive model that can be used to assess the potential of a given WTR for P control. The model input would only require typical chemical parameters to characterize the WTR eliminating the need for further laboratory analysis of the residuals.
- Defining clear and simple recommendations for handling, storing, and transporting WTR that will minimize management costs while maximizing beneficial P control characteristics.
- $CaCl_2$ extractable P has been shown to represent the release of P within a soil solution and thus, P that may pass through a soil in soil solution. This weak ionic extractant was developed to buffer the differing ionic strengths of soil solutions. This was the basis for selection of this method in the research for this study. Water extractable soil P is becoming widely accepted as representing the release of soil P to rainfall and runoff water, and thus, the transfer of soil P to runoff (McDowell and Sharpley 2001). Also, water extractable P values are greater than $CaCl_2$-P and may give more sensitivity among samples. Finally, water extraction is gaining acceptance as an environmental P test for manures and biosolids to estimate their relative potentials to P in runoff water. Therefore, soluble P via water extraction should be compared to $CaCl_2$ extraction for WTR and soil/organic byproduct blends to determine which method best relates to the impact of WTR on bioavailable P.
- This project focused on P loss due to the magnitude of impact that P loss has on surface water quality. There may be other constituents worthy of measurement in runoff waters that would address other potential public health concerns, including but not limited to nitrogen, arsenic, chromium, copper, and zinc. Although these losses may be small, the perceived impacts by the public or regulating agencies may be limiting to their use. Therefore, future research efforts of field scale demonstration should include monitoring for constituents of concern. Determining which constituents are of concern requires focused dialogues with regulating agencies.
- While WTR can bind P in-situ or in runoff water, the fate of WTR bound P that enters the aquatic environment with particulates due to erosion should be evaluated.
- Plant uptake or accumulation of other WTR constituents via plant tissue analyses and crop yields should be documented at field scale trials to further assess the impact of WTR on the soil matrix and plants.

- The longevity of the effect of WTRs on soils also should be considered to determine if WTRs have to be reapplied after a certain number of years to accomplish the same effect on soil P. Such evaluation would include an assessment of the transformation of P between inorganic and organic forms.
- This research was conducted with one soil type to limit the variables for determining the impact of WTR on both soil test and soluble P. Earlier studies (DeWolfe 1990, Sharpley, McDowell and Kleinman 2004) have shown that the impacts of WTR vary as a function of soil type and on the dominant form and chemistry of P in the soil. Additional studies on other soils types also are recommended to identify which soil parameters make the greatest difference in the ability of WTR to have both short term and long term effects on soil P.
- The impact of WTR on soil biotic properties (i.e., soil fungi) should be examined, since P transformation can be substantially impacted by these properties (Yost and Fox 1979).

APPENDIX A
DATABASE FORM USED TO RECORD WTP DATA

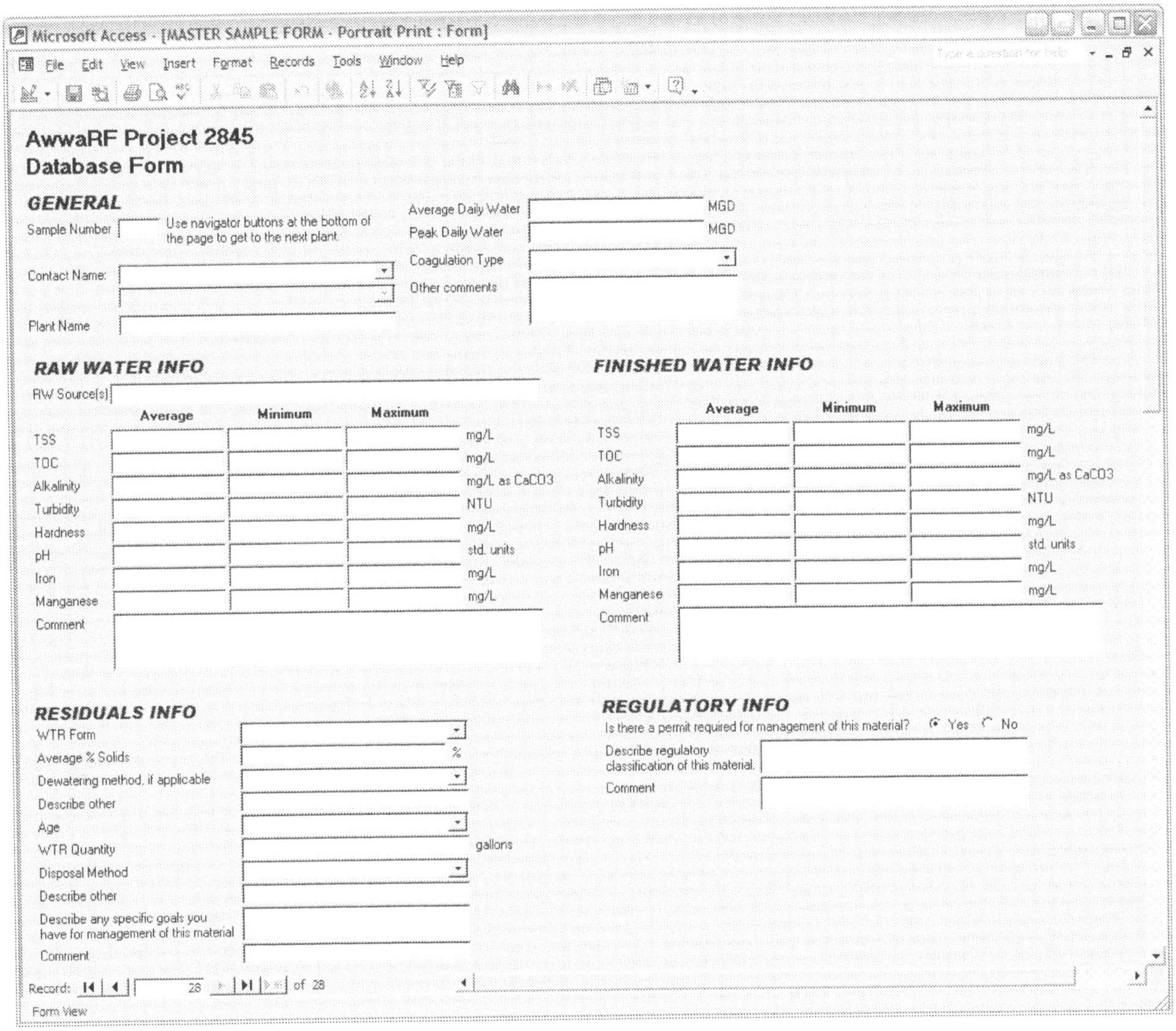
Microsoft Access - [MASTER SAMPLE FORM - Portrait Print : Form]

File Edit View Insert Format Records Tools Window Help

AwwaRF Project 2845
Database Form

GENERAL

Sample Number — Use navigator buttons at the bottom of the page to get to the next plant.

Contact Name:

Plant Name

Average Daily Water — MGD

Peak Daily Water — MGD

Coagulation Type

Other comments

RAW WATER INFO

RW Source(s)

	Average	Minimum	Maximum	
TSS				mg/L
TOC				mg/L
Alkalinity				mg/L as CaCO3
Turbidity				NTU
Hardness				mg/L
pH				std. units
Iron				mg/L
Manganese				mg/L

Comment

FINISHED WATER INFO

	Average	Minimum	Maximum	
TSS				mg/L
TOC				mg/L
Alkalinity				mg/L as CaCO3
Turbidity				NTU
Hardness				mg/L
pH				std. units
Iron				mg/L
Manganese				mg/L

Comment

RESIDUALS INFO

WTR Form

Average % Solids — %

Dewatering method, if applicable

Describe other

Age

WTR Quantity — gallons

Disposal Method

Describe other

Describe any specific goals you have for management of this material

Comment

REGULATORY INFO

Is there a permit required for management of this material? Yes No

Describe regulatory classification of this material.

Comment

Record: 28 of 28

Form View

(continued)

DATABASE FORM USED TO RECORD WTP DATA (CONTINUED)

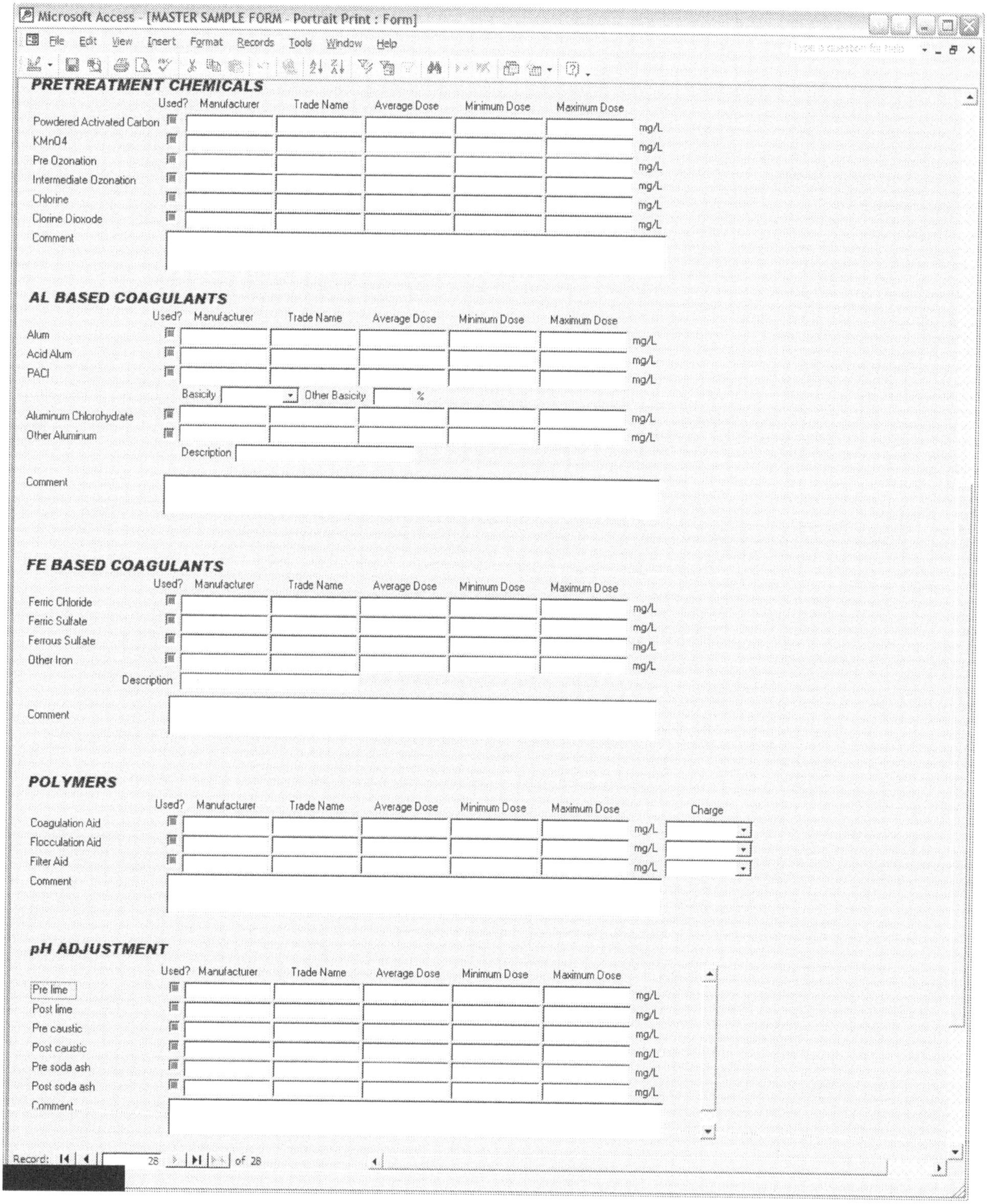

Microsoft Access - [MASTER SAMPLE FORM - Portrait Print : Form]

File Edit View Insert Format Records Tools Window Help

PRETREATMENT CHEMICALS

	Used?	Manufacturer	Trade Name	Average Dose	Minimum Dose	Maximum Dose	
Powdered Activated Carbon							mg/L
KMnO4							mg/L
Pre Ozonation							mg/L
Intermediate Ozonation							mg/L
Chlorine							mg/L
Clorine Dioxode							mg/L

Comment

AL BASED COAGULANTS

	Used?	Manufacturer	Trade Name	Average Dose	Minimum Dose	Maximum Dose	
Alum							mg/L
Acid Alum							mg/L
PACl							mg/L
		Basicity		Other Basicity	%		
Aluminum Chlorohydrate							mg/L
Other Aluminum							mg/L

Description

Comment

FE BASED COAGULANTS

	Used?	Manufacturer	Trade Name	Average Dose	Minimum Dose	Maximum Dose	
Ferric Chloride							mg/L
Ferric Sulfate							mg/L
Ferrous Sulfate							mg/L
Other Iron							mg/L

Description

Comment

POLYMERS

	Used?	Manufacturer	Trade Name	Average Dose	Minimum Dose	Maximum Dose		Charge
Coagulation Aid							mg/L	
Flocculation Aid							mg/L	
Filter Aid							mg/L	

Comment

pH ADJUSTMENT

	Used?	Manufacturer	Trade Name	Average Dose	Minimum Dose	Maximum Dose	
Pre lime							mg/L
Post lime							mg/L
Pre caustic							mg/L
Post caustic							mg/L
Pre soda ash							mg/L
Post soda ash							mg/L

Comment

Record: 28 of 28

APPENDIX B
MANURE ANALYSIS REPORT FOR POULTRY MANURE

PENNSTATE

1855

(814) 863-0841 Fax (814) 863-4540

Agricultural Analytical Services Laboratory
The Pennsylvania State University
University Park PA 16802

Analysis Report For:	Copy To:
Jim DeWolfe 712 Hampshire Circle State College Pa 16803	

LAB ID:	SAMPLE ID:	REPORT DATE:	DATE SAMPLED:	COUNTY:	MATERIAL:	TYPE:	STORAGE SYSTEM:
M02865	OK-CL	5/27/2003	5/13/03	Centre	Manure	Poultry (Broilers)	Bedded Pack or Litter

MANURE ANALYSIS REPORT

Results on as sampled (wet weight) basis

Analyte	lb/ton	lb/1000 gal	Analyte	lb/ton	lb/1000 gal
Solids: 73.61 %			Total Calcium (Ca)	52.77	220.05
Total Nitrogen (N)	53.82	224.44	Total Magnesium (Mg)	9.31	38.83
Ammonium N (NH_4-N)	12.79	53.32	Total Sulfur (S)	8.47	35.32
Calculated Organic N	41.04	171.12	Total Copper (Cu)	0.61	2.56
Total Phosphate (P_2O_5)	63.56	265.05	Total Zinc (Zn)	0.57	2.39
Total Potash (K_2O)	42.26	176.24	Total Manganese (Mn)	0.72	3.00
			Total Iron (Fe)	6.81	28.42
			Total Sodium (Na)	8.93	37.25
			Total Aluminum (Al)	11.13	46.42

Optional Test Results:	pH	Carbon (C) (%)	C:N Ratio	Ash (%)	Volatiles (%)	Nitrate Nitrogen (lb/ton)	Nitrate Nitrogen (lb/1000 gal)

Comments:

- The enclosed fact sheet "Using Your Manure Analysis Report" provides information to help you interpret this report and calculate appropriate manure application rates for your crops.
- Manure nutrients are not all equivalent to fertilizer nutrients. Phosphorus and potassium can be substituted directly for fertilizer to meet your soil test recommendation. Nitrogen (N) availability varies with handling. This must be accounted for in utilizing manure to meet soil test N recommendations. See the enclosed fact sheet "Using Your Manure Analysis Report"

REFERENCES

Alloway, B.J. 1990. *Heavy Metals in Soils*. New York: John Wiley and Sons, Inc.

Bache, B.W. and E.G. Williams. 1971. A Phosphate Sorption Index for Soils. *Journal of Soil Science*, 22 (3) 289-301.

Basta, N.T., and D.E. Storm. 1997. Use of Alum Residuals to Reduce Nutrient Runoff From Agricultural Land Treated With Animal Manure to Protect Surface Water Quality. In *Proc. of WEF/AWWA Joint Residuals and Biosolids Management Conference.* Denver, Colo.: AWWA.

Basta, N.T., E.A. Dayton, and L.E. Gallimore. 1999. Nutrient Adsorption Capacity of Water Treatment Residuals. In *Proc. of WEF/AWWA Joint Residuals and Biosolids Management Conference.* Denver, Colo.: AWWA.

Chesapeake Bay Program. 2004. Reducing Nutrient Pollution [Online]. Available: <http://www.chesapeakebay.net/nutr2.htm>. [cited January 8, 2005]

Colorado Department of Public Health and Environment, Water Quality Control Commission. 2003. 5 CCR 1002-64, Biosolids Regulation, Denver, CO.

Colorado Department of Public Health and Environment, Water Quality Control Division. 2004. 5 CCR 1003-7, Regulations Pertaining to the Beneficial Use of Water Treatment Sludge and Fees Applicable to the Beneficial Use of Sludges, Denver, CO:

Cornwell, D.A., M.M. Bishop, R.G. Gould and C/ Vandermeyden. 1987. *Water Treatment Plant Waste Management*. Denver, Colo.: AwwaRF and AWWA.

Cornwell, D.A., R.N. Mutter, and C. Vandermeyden. 2000. *Commercial Application and Marketing of Water Plant Residuals*. Denver, Colo.: AwwaRF and AWWA.

Dahnke W.C. and D.A. Whitney. 1988. *Measurement of Soil Salinity*. In: Dahnke W.C., ed. Recommended chemical soil test procedures for the north central region. Fargo, ND: North Dakota State Univ, 1988:32-33 North Cen. Regional North Dakota Agric. Exp. Stn..

Daniel, T.C., G. Hausten, P. Moore, and Nichols, D. 1999. Effect of Land Applied Alum Residuals and HCA on Runoff Phosphorus Concentrations From High-P Soils. In *Proc. of WEF/AWWA Joint Residuals and Biosolids Management Conference.* Denver, Colo.: AWWA.

Dayton, E.A., N.T. Basta, and C.A. Jakober. 2001. Phosphorus Sorption Capacity of Drinking Water Treatment Residuals and Potential for Use to Protect Water Quality. In *Proc. of WEF/AWWA Joint Residuals and Biosolids Management Conference.* Denver, Colo.: AWWA.

DeWolfe, J. R. 1990. The Effects of Land Applied Alum and Ferric Chloride Water Treatment Sludges on Soil Phosphorus. Masters Thesis, Department of Civil Engineering, University Park, PA: The Pennsylvania State University.

DeWolfe, J.R. and Brandt, R.C., 1997. Beneficial Use of Water Treatment Residual as a Soil Amendment. In *Proc. of WEF/AWWA Joint Residuals and Biosolids Management Conference.* Denver, Colo.: AWWA.

Eaton, R.A. and J.T. Sims. 2001. Effect of Water Treatment Residuals on Growth and Phosphorus Removal by Plants Adapted for Vegetative Buffer Strips in the Mid-Atlantic U.S. In *Proc. of WEF/AWWA Joint Residuals and Biosolids Management Conference.* Denver, Colo.: AWWA.

Elliott, H.A., Dempsey, B.A., Hamilton, D.W., and DeWolfe, J.R. 1990. *Land Application of Water Treatment Sludges: Impact and Management.* Denver, Colo.: AwwaRF and AWWA.

Gallimore, L.E.; Basta, N.T.; Storm, D.E.; Payton, M.E.; Huhnke, R.H.; & M.D. Smolen. 1999. Use of Water Treatment Residuals to Reduce Nutrients in Surface Runoff From Agricultural Land. *J. Environ. Qual.* 28:1474-1478.

Ippolito, J.A.; Barbarick, K.A.; & Redente, E.F. 1999. Co-Application Effects of Water Treatment Residuals and Biosolids on Two Range Grasses. *J. Environ. Qual.* 28:1664-1650.

Ippolito, J.A., and Ken Barbarick. 2001. Environmental Management of Biosolids and Water Treatment Residuals. In *Proc. of WEF/AWWA Joint Residuals and Biosolids Management Conference.* Denver, Colo.: AWWA.

Ippolito, J.A, K. Barbarick, D. Heil, J. Chandler and E. Redente. 2003. Phosphorus Retention Mechanisms of a Water Treatment Residual, *J. Env. Quality*, 32:1577-1582.

Jacobs, L.W., and B.J. Teppen. 2001. Alum WTR as an Amendment for Phosphorus-Enriched Soils. p.52. In WEF/AWWA Joint Residuals and Biosolids Management Conference Abstracts, Feb. 21-24, San Diego, CA.

Jokela, W.E. 2000. A Phosphorus Index for Vermont. 2000. In *Proc. from Managing Nutrients and Pathogens from Agriculture.* Camp Hill, PA. Ithaca, NY: NRAES.

Kleinman, P.J.A. and Sharpley, A.N. 2002. Estimating Soil Phosphorus Sorption Data from Mehlich-3 Data. *Commun. Soil Sci. Plant Anal.*, 33(11 & 12), 1825-1839.

Kuo S. 1996. Phosphorus. pp. 869-919. in D.L. Sparks et al., (eds) *Methods of Soil Analysis, Part 3, Chemical Methods*. Madison, WI: ASA, CSSA and SSSA.

Leeds, R., L. C. Brown, N. L. Watermeier. 1992. Nonpoint Source Pollution: Water Primer. Ohio State University Extension Fact Sheet AEX-465-93. Columbus, OH: OSU.

Lemunyon, J.L. and R.G. Gilbert. 1993. The Concept and Need for a Phosphorus Assessment Tool. *J. Prod. Agric.* 6:483-496.

MacPhee, M.J., D.A. Cornwell and R. Brown. 2002. *Trace Contaminants in Drinking Water Chemicals*. Denver, Colo. AwwaRF and AWWA.

McDowell, R.W. and Sharpley, A.N. 2001. Approximating Phosphorus Release from Soils to Surface Runoff and Subsurface Drainage. *J. Environ. Qual.* 30:508–520

McKeague, J.A, and J.H. Day. 1966. Ammonium Oxalate Extraction of Amorphous Iron and Aluminum. In M.R. Carter (ed.). *Soil Sampling and Methods of Analysis.* Canadian Society of Soil Science. Boca Raton, LA: Lewis Publishers.

Mehlich, A. 1984. Mehlich-3 soil test extractant: A modification of Mehlich-2 extractant. *Commun. Soil Sci. Plant Anal.* 15:1409.

Miller, W.P. 1987. A Solenoid-operated, Variable Intensity Rainfall Simulator. *J. Soil Sci. Soc. Am.* 51:832-834.

Miller, D.M., T. Alvisyahrin, T.C. Daniel, and P.A. Moore, Jr., 1997. Characterization and Use of Alum-Based Water Treatment Residuals to Reduce Water-Soluble Phosphorus on Excessively Manured Soils. In Proc. of the Arkansas WRC annual Research Conference. AWRC Publication No. MSC-210. Fayetteville, AR: Univ. of Arkansas.

Morris, T.F and J.E. Hyde. 2001. Effect of Unamended and Phosphorus-Amended Water Treatment Residual on Corn Yields. p. 54. In WEF/AWWA Joint Residuals and Biosolids Management Conference Abstracts, Feb. 21-24, San Diego, CA.

Mulvaney, R.L. 1996.Nitrogen–Inorganic Forms. pp. 1130-1131, 1155-1159. in D.L. Sparks et al., (eds) *Methods of Soil Analysis, Part 3, Chemical Methods*. ASA, CSSA and SSSA, Madison, WI: ASA, CSSA, SSSA.

Novak, J.M., D.W. Watts, P.G. Hunt, and K.C. Stone. 2000. Phosphorus Movement Through a Coastal Plain Soil After a Decade of Intensive Swine Manure Application. *J. Environ. Qual.* 29:1310-11315.

Novak, J.M, P.G. Hunt, K.C. Stone, D.W. Watts and M.H. Johnson. 2002. Riparian Zone Impact on Phosphorus Movement to a Coastal Plain Black Water Stream. *J. Soil and Water Conservation*, 57(3):127-133.

Novak, J.M., and D.W. Watts. 2004. Increasing the Phosphorus Sorption Capacity of Southeastern Coastal Plain Soils Using Water Treatment Residuals. *J. Soil Science*, 169(3):206-214.

O'Connor, G.A. and H.A. Elliott. 2000. *Co-Application of Biosolids and Water Treatment Residuals*. Final Report. Tallahassee, FL: Florida Dept. of Environmental Protection.

Pennsylvania Association of Conservation Districts. 2003. Mini-Grant Programs [Online]. Available: <http://www.pacd.org/products/mini_programs.htm>. [cited January 8, 2005]

Peters, J.M. and N.T. Basta. 1996. Reduction of Excessive Bioavailable Phosphorus in Soils by Using Municipal and Industrial Wastes. *J. Environ. Qual.* 25:1236-1241.

Rengasamy, P., J.M. Oades, T.W. Hancock. 1980. The Improvement of Soil Structure and Plant Growth by Addition of Alum Sludge. *Communications in Soil Science and Plant Analysis*, 11(6):533-545.

Rhoades J.D. 1996. Salinity: Electrical Conductivity and Total Dissolved Solids P. 420. in D.L. Sparks et al., (eds) *Methods of Soil Analysis, Part 3, Chemical Methods*. Madison, WI: ASA, CSSA and SSSA.

Sharkoff, J.L, R.M Waskom, and J.G. Davis. 2003. *Colorado Phosphorus Index Risk Assessment: Version 3.0*. USDA-NRCS Agronomy Technical Note CO-95. Lakewood, Colo.: USDA.

Sharpley A.N. and J. Syers. 1979. Loss of Nitrogen and Phosphorus in Tile Drainage as Influenced by Urea Application and Grazing Animals. *New Zealand Journal of Agricultural Research,* 22:127-131.

Sharpley, A.N., T.C. Daniel, J.T. Simms, and D.H. Pote. 1996. Determining Environmentally Sound Soil Phosphorus Levels. *J. Soil Water Conservation*, 51:160-166.

Sharpley, A.N., ed. 2000. *Agriculture and Phosphorus Management – The Chesapeake Bay*, Boca Raton, FL: CRC Press.

Sharpley, A.N. and D. Beegle. 2001. Managing Phosphorus for Agriculture and the Environment. Catalog No. UC162. *The Pennsylvania State University College of Agricultural Sciences*. University Park, PA.

Sharpley, A.N., J. Weld, D. Beegle, P. Kleinman, W. Gburek, P. Moore, G. Mullins. 2003. Development of Phosphorus Indices for Nutrient Management Planning Strategies in the United States. *J. of Soil and Water Conservation*, 58(3):137-152.

Sharpley, A.N., McDowell, R.W., and Kleinman, P.J.A. 2004. Amounts, Forms, and Solubility of Phosphorus in Soils Receiving Manure, *J. Soil Science. Soc. Am.* 68:2048–2057.

Sims, J.T., N.J. Luka, and R.A. Eaton. 2001. Can Water Treatment Residuals be Integrated Into Best Management Practices to Reduce Nonpoint Source Pollution of the Chesapeake Bay by Agricultural Phosphorus? In *Proc. of WEF/AWWA Joint Residuals and Biosolids Management Conference.* Denver, Colo.: AWWA.

Sims, J.T., R. Simard, and B. Joern. 1998. Phosphorus Loss in Agricultural Drainage: Historical Perspective and Current Research. *J. Environ. Qual.* 27:277-293.

Street, J.R., M. Watson and W. Pound. 2004. *Interpreting a Soil Test for Lawns*, Ohio State University Extension Pub. HYG-4028, Columbus, OH: The Ohio State University.

Thomas, G.W. 1996. Soil pH and Soil Acidity P. 487. in D.L. Sparks et al., (eds) *Methods of Soil Analysis, Part 3, Chemical Methods*. ASA, CSSA and SSSA, Madison, WI

U.S. Department of Agriculture National Resource Conservation Service. 2004. Oklahoma Conservation Practice Code 590 - Nutrient Management. Available: <http://efotg.nrcs.usda.gov/treemenuFS.aspx?Fips=40041&MenuName=menuOK.zip> [cited December 20, 2004]

U.S. Department of Agriculture National Resource Conservation Service. 2003. Conservation Practice Code 590 - Nutrient Management. Available: <http://www.nrcs.usda.gov/technical/standards/nhcp.html>. [cited December 20, 2004]

U.S. Department of Agriculture National Resource Conservation Service. 1994. The Phosphorus Index: A Phosphorus Assessment Tool [Online]. Available: <http://www.nrcs.usda.gov/technical/ECS/nutrient/pindex.html>. [cited December 20, 2004]

U.S. Department of Agriculture National Resource Conservation Service. 1999. *Animal Waste Management*, National Engineering Handbook Part 651, Agricultural Waste Management Field Handbook. Washington, D.C.: U.S. Government Printing Office,

U.S. Department of Agriculture and U.S. Environmental Protection Agency. 1999. Unified National Strategy for Animal Feeding Operations. Washington, D.C.: U.S. Government Printing Office.

U.S. Environmental Protection Agency. 1992. Method 1311, Toxicity Characteristic Leaching Procedure [Online]. Available: <http://www.epa.gov/epaoswer/hazwaste/test/pdfs/1311.pdf>. [cited January 6, 2003]

U.S. Environmental Protection Agency. 1994. Method 3051. Microwave Assisted Acid

U.S. Environmental Protection Agency. 1997. State Source Water Assessment and Protection Programs (United States Environmental Protection Agency – Office of Water), pub. EPA 816-R-97-009, Washington, D.C.: USEPA.

U.S. Environmental Protection Agency . 1998. 40 CFR Parts 9, 141 and 142. National Primary Drinking Water Regulations: Disinfe ctants and Disinfection Byproducts; Final Rule (Stage 1 Disinfectants and Disinfection Byproducts Rule (Stage 1 DBPR)). *Fed. Reg.*(Part II), 63(241): 69390 – 69476.

U.S. Environmental Protection Agency. 2000. 40 CFR Parts 9, 141 and 142. National Primary Drinking Water Regulations; Radionuclides; Final Rule. *Fed. Reg.* (Part II), 65(236): 76708-76753.

U.S. Environmental Protection Agency. 2003. 40 CFR Parts 141 and 142. National Primary Drinking Water Regulations: Long Term 2 Enhanced Surface Water Treatment Rule; Proposed Rule. *Fed. Reg.* (Part II), 68(154): 47640-47795.

Weld, J.L., D. Beegle, W. Gburek, P. Kleinman, A. Sharpley. 2003. The Pennsylvania Phosphorus Index: Version 1. Published for USDA-ARS, University Park, PA.: The Pennsylvania State University,

Yost, R.S. and R.L. Fox. 1979. Contribution of Mycorrhizae to Phosphorus Nutrition of Crops Growing in an Oxisol. *Agronomy Journal*, 71:903-908.

ABBREVIATIONS

Al	aluminum
Al-ox	reactive aluminum oxide
alum	aluminum sulfate
Agric.	Agriculture
Am.	America
AOE	ammonium oxalate extractable
ARS	Agricultural Research Service
As	arsenic
ASA	American Society of Agronomists
AWWA	American Water Works Association
AwwaRF	Awwa Research Foundation
BMP	best management practice
$CaCl_2$	calcium chloride
CAFO	concentrated animal feeding operation
CCE	calcium carbonate equivalence
CEC	cation exchange capacity
CDPHE	Colorado Department of Public Health and Environment
CNMP	comprehensive nutrient management plan
COPI	Colorado P Index
CSSA	Crop Science Society of America
CSREES	Cooperative States Research, Education, and Extension Service
CWA	Clean Water Act
DW	Denver Water
EPA	Environmental Protection Agency
Fe	iron
Fe-ox	reactive iron oxide
Fed. Reg.	Federal Register
GAC	granular activated carbon
gpm	gallons per minute
IESWTR	Interim Enhanced Surface Water Treatment Rule
ICP-AES	inductively coupled plasma atomic emission spectroscopy
J.	Journal
Jour.	Journal
$KMnO_4$	potassium permanganate

LT2ESWTR	Long Term 2 Enhanced Surface Water Treatment Rule
M-DBP	microbial and disinfection byproduct
MCL	maximum contaminant level
MCLG	maximum contaminant level goal
mg/kg	milligrams per kilogram
Mg/ha	megagrams per hectare
mg/L	milligrams per liter
mgd	million gallons per day
Mn	manganese
N	nitrogen
No.	number
NPS	non-point source
NRAES	Natural Resource, Agriculture, and Engineering Service
NRCS	National Resource Conservation Service
NTU	nephelometric turbidity units
P	phosphorus
P_2O_5	phosphorus oxide
PA DEP	Pennsylvania Department of Environmental Protection
PAC	Project Advisory Committee
PAC	powdered activated carbon
PACD	Pennsylvania Association of Conservation Districts
PACl	polyaluminum chloride
PAWC	Pennsylvania American Water Company
Pmax	inorganic-P sorption maximum value
ppm	parts per million
Proc.	Proceedings
Q	flow rate in gallons per minute
QA	quality assurance
RUSLE	revised universal soil loss equation
S	furrow slope in feet per 100 feet, expressed in percent
Sci.	Science
SDWA	Safe Drinking Water Act
SERA	Southern Extension Research Activity
Soc.	Society
SSSA	Soil Science Society of America.
Stage 1 DBPR	Stage 1 Disinfectants and Disinfection Byproducts Rule
T&O	taste and odor
TOC	total organic carbon

TMDL	Total Maximum Daily Load
TMUA	Tulsa Metropolitan Utility Authority
U.S.	United States
USA	United States of America
USDA	United States Department of Agriculture
WEF	Water Environment Federation
WTP	water treatment plant
WTR	water treatment residual

IWA Publishing's authorised EU representative for General Product Safety Regulations is Diane D'Arras, 15 rue Duret, 75116 Paris, France, e-mail: safety@iwap.co.uk.

Printed and bound by CPI Group (UK) Ltd, Croydon, CR0 4YY
12/05/2026
02108886-0010